TRANSPORTATION INFRASTRUCTURE – ROADS, HIGHWAYS, BRIDGES, AIRPORTS AND MASS TRANSIT

U.S. BRIDGE CONDITIONS AND LONG-TERM BRIDGE PERFORMANCE

A PRIMER

TRANSPORTATION INFRASTRUCTURE – ROADS, HIGHWAYS, BRIDGES, AIRPORTS AND MASS TRANSIT

Additional books in this series can be found on Nova's website under the Series tab.

Additional E-books in this series can be found on Nova's website under the E-book tab.

TRANSPORTATION INFRASTRUCTURE – ROADS, HIGHWAYS, BRIDGES, AIRPORTS AND MASS TRANSIT

U.S. BRIDGE CONDITIONS AND LONG-TERM BRIDGE PERFORMANCE

A PRIMER

RUBY L. LACEY
EDITOR

New York

For permission to use material from this book please contact us:
Telephone 631-231-7269; Fax 631-231-8175
Web Site: http://www.novapublishers.com

Library of Congress Cataloging-in-Publication Data

ISBN: 978-1-63117-486-5

Published by Nova Science Publishers, Inc. † New York

CONTENTS

PREFACE

The United States has approximately 607,000 bridges on public roads subject to the National Bridge Inspection Standards mandated by Congress. The sudden catastrophic failure of the I-5 Interstate System bridge in Washington State has raised policy concerns in Congress regarding the condition of the nation's transportation infrastructure in general, and in particular the federal role in funding, building, maintaining, and ensuring the safety of roads and especially bridges in the United States. This book discusses highway bridge conditions.It then continues to provide information on a Long-Term Bridge Performance (LTBP) Program that is intended to provide a comprehensive definition of bridge performance that will be the foundation for carefully designed research studies in the LTBP Program.

Chapter 1 – The sudden catastrophic failure of the I-5 Interstate System bridge in Washington State on May 23, 2013, has raised policy concerns in Congress regarding the condition of the nation's transportation infrastructure in general, and in particular the federal role in funding, building, maintaining, and ensuring the safety of roads and especially bridges in the United States. Of the 607,000 public road bridges, about 67,000 (11%) were classified as structurally deficient in 2012, and another 85,000 (14%) were classified as functionally obsolete. This is less than half the number classified as structurally deficient in 1990 and 16% less than were classified as functionally obsolete. Structurally deficient and functionally obsolete bridges are not necessarily unsafe. Nonetheless, public concern about bridge safety in the wake of the I-5 bridge collapse raises the policy question of how quickly these bridges should be replaced or improved. At current annual spending levels, the Federal Highway Administration (FHWA) estimates that the bridge investment backlog (in dollar terms) would be reduced by 11% by 2028. Reducing the backlog to near zero during the same period is estimated to require an annual spending rate roughly 60% higher than recent levels. The most recent highway bill, the Moving Ahead for Progress in the 21st Century Act (MAP-21; P.L. 112-141), eliminated the former Highway Bridge Program, which distributed federal money specifically for bridge improvements. States may use funds received under two major FHWA programs, the National Highway Performance Program and the Surface Transportation Program, for bridge repairs or construction, but the decision about how much of its funding to devote to bridges rather than roadway needs is up to each state. FHWA enforces certain planning requirements and performance standards established in MAP-21, but it does not make the determination as to which bridges should benefit from federal funding. Congressional issues regarding the nation's highway bridge infrastructure include the following: Given the steady decline in the number of structurally deficient bridges during recent decades, should Congress take action to accelerate the improvement of the remaining deficient bridges? If Congress wishes to accelerate the reduction in the number of deficient bridges under MAP-21, what can it do to encourage the states to spend more of their federal funds on their deficient bridges? Given the context of large projected shortfalls in highway trust fund revenues relative to spending, should Congress encourage increased spending on highway bridges through increased use of tolling and public private partnerships (PPPs)? Should Congress consider legislation to redirect spending away from off-system bridges to more heavily used bridges on the designated federal-aid highways? Congressional oversight of bridge conditions could be complicated by the absence of a freestanding program. How quickly can FHWA develop the MAP-21 performance measures to report to Congress on progress on bridge conditions?

Chapter 2 – The performance of bridges is critical to the overall performance of the highway transportation system in the United States. However, many critical aspects of bridge performance are not well understood. The reasons for this include the extreme diversity of the bridge infrastructure, the widely varying conditions under which bridges serve, and the lack of reliable data needed to understand performance. The Long-Term Bridge Performance (LTBP) Program was created to identify, collect, and analyze research-quality data on the most critical aspects of bridge performance. This report describes the bridge infrastructure in the United States and explains why reliable indicators of bridge performance are necessary. Current methods of measuring performance are described, and the need for more precisely targeted performance indicators is discussed. The report explains bridge performance in terms of the aspects that have the most impact on the ability of the bridge to serve its intended purpose in a safe, efficient, and economical manner. The purpose of this report is to provide an understanding of what bridge performance means and the factors that influence bridge performance. It describes the various ways in which bridge owners measure and report bridge performance. It also describes how the LTBP Program will create new measures to allow deeper analysis and understanding, which will lead to ways to improve long-term bridge performance.

In: U.S. Bridge Conditions …
Editor: Ruby L. Lacey

ISBN: 978-1-63117-486-5

Chapter 1

HIGHWAY BRIDGE CONDITIONS: ISSUES FOR CONGRESS*

Robert S. Kirk and William J. Mallett

SUMMARY

The sudden catastrophic failure of the I-5 Interstate System bridge in Washington State on May 23, 2013, has raised policy concerns in Congress regarding the condition of the nation's transportation infrastructure in general, and in particular the federal role in funding, building, maintaining, and ensuring the safety of roads and especially bridges in the United States.

Of the 607,000 public road bridges, about 67,000 (11%) were classified as structurally deficient in 2012, and another 85,000 (14%) were classified as functionally obsolete. This is less than half the number classified as structurally deficient in 1990 and 16% less than were classified as functionally obsolete. Structurally deficient and functionally obsolete bridges are not necessarily unsafe. Nonetheless, public concern about bridge safety in the wake of the I-5 bridge collapse raises the policy question of how quickly these bridges should be replaced or improved. At current annual spending levels, the Federal Highway Administration (FHWA) estimates that the bridge investment backlog (in dollar terms) would be reduced by 11% by 2028. Reducing the backlog to near zero

* This is an edited, reformatted and augmented version of a Congressional Research Service publication, CRS Report for Congress R43103, prepared for Members and Committees of Congress, from www.crs.gov, dated December 19, 2013.

during the same period is estimated to require an annual spending rate roughly 60% higher than recent levels.

The most recent highway bill, the Moving Ahead for Progress in the 21st Century Act (MAP-21; P.L. 112-141), eliminated the former Highway Bridge Program, which distributed federal money specifically for bridge improvements. States may use funds received under two major FHWA programs, the National Highway Performance Program and the Surface Transportation Program, for bridge repairs or construction, but the decision about how much of its funding to devote to bridges rather than roadway needs is up to each state. FHWA enforces certain planning requirements and performance standards established in MAP-21, but it does not make the determination as to which bridges should benefit from federal funding.

Congressional issues regarding the nation's highway bridge infrastructure include the following:

- Given the steady decline in the number of structurally deficient bridges during recent decades, should Congress take action to accelerate the improvement of the remaining deficient bridges?
- If Congress wishes to accelerate the reduction in the number of deficient bridges under MAP-21, what can it do to encourage the states to spend more of their federal funds on their deficient bridges?
- Given the context of large projected shortfalls in highway trust fund revenues relative to spending, should Congress encourage increased spending on highway bridges through increased use of tolling and public private partnerships (PPPs)?
- Should Congress consider legislation to redirect spending away from off-system bridges to more heavily used bridges on the designated federal-aid highways?
- Congressional oversight of bridge conditions could be complicated by the absence of a freestanding program. How quickly can FHWA develop the MAP-21 performance measures to report to Congress on progress on bridge conditions?

BACKGROUND

The United States has approximately 607,000 bridges on public roads subject to the National Bridge Inspection Standards mandated by Congress.[1] About 48% of these bridges are owned by state governments and 50% by local governments. State governments generally own the larger and more heavily traveled bridges, such as those on the Interstate Highway system. Only 1.5%

of highway bridges are owned by the federal government, primarily those on federally owned land.

About 9% of all bridges carry Interstate Highways, and another 25% serve arterial highways other than Interstates.[2] Interstate and other major arterial bridges carry almost 80% of average daily traffic. The highest traffic loads are on Interstate Highway bridges in urban areas; these account for only 5% of all bridges, but carried 36% of average daily traffic in 2012.[3]

Bridge Conditions

Federal law requires states to periodically inspect public road bridges and to report these findings to the Federal Highway Administration (FHWA). This information permits FHWA to characterize the existing condition of a bridge compared with one newly built and to identify those that are structurally deficient or functionally obsolete. A bridge is considered structurally deficient "if significant load-carrying elements are found to be in poor or worse condition due to deterioration and/or damage, or the adequacy of the waterway opening provided by the bridge is determined to be extremely insufficient to the point of causing intolerable traffic interruptions."[4]

A functionally obsolete bridge, on the other hand, is one whose current geometric characteristics—deck geometry (such as the number and width of lanes), roadway approach alignment, and over/underclearances—do not meet current design standards or traffic demands. A bridge can be both structurally deficient and functionally obsolete, but structural deficiencies take precedence in classification. As a result, a bridge that is structurally deficient and functionally obsolete is classified in the FHWA's National Bridge Inventory as structurally deficient.

A bridge classified as structurally deficient or functionally obsolete is not necessarily unsafe, but may require the posting of a vehicle weight or height restriction.

The proportion of bridges classified as structurally deficient has declined 54% since 1990, and fell every year between 1990 and 2012 (see *Figure 1*). In 2012, approximately 67,000 bridges, or 11% of the total number of bridges, were classified as structurally deficient, as compared to 138,000 in 1990. The number of functionally obsolete bridges declined by 16% over the same period.

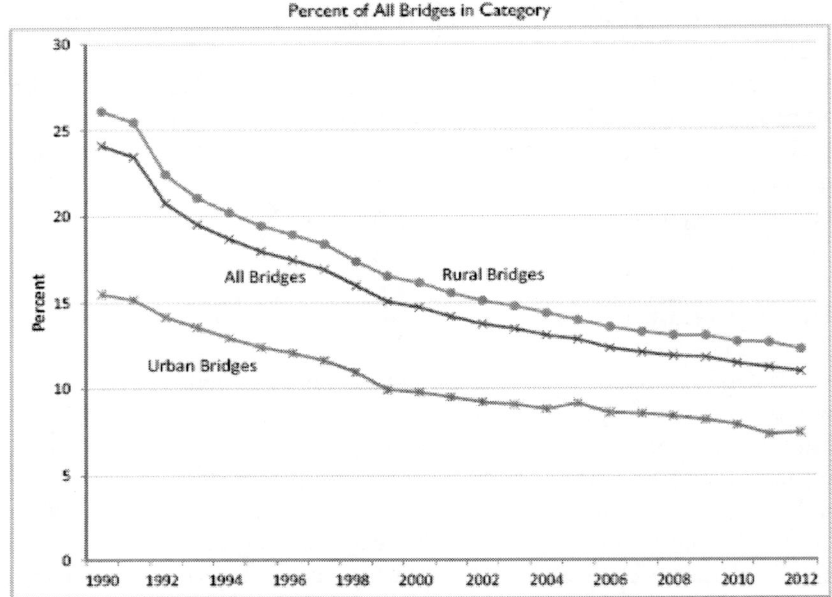

Source: 1990-2010: U.S. Department of Transportation, Research and Innovative Technology Administration, Bureau of Transportation Statistics, National Transportation Statistics, table 1-28; 2011-2012: U.S. Department of Transportation, Federal Highway Administration, National Bridge Inventory, "Count, Area, and Length of Bridges by Highway System.

Figure 1. Structurally Deficient Bridges in the United States, 1990-2012.

Bridges on the most heavily traveled roads, such as Interstates and other arterials, are generally in better condition than bridges on more lightly traveled routes. For example, 4.7% of urban interstate bridges were considered structurally deficient in 2012, about half of the 9.8% structural deficiency rate of urban bridges on local roads. Likewise, 4.1% of rural Interstate Highway bridges were structurally deficient in 2012, about a quarter of the 17.2% structural deficiency rate of bridges on rural roads handling local traffic.

As the bridges on local roads are usually owned by local governments, locally owned bridges had more than twice the structural deficiency rate of state-owned bridges in 2012. Some 14.8% of locally owned bridges were categorized as structurally deficient in 2012, versus 7.0% of state-owned bridges. For bridge deficiency and obsolescence rates by state see *Appendix A*.

Future Bridge Funding Needs

Every two years, FHWA assesses the condition and performance of the nation's highways and bridges, documents current spending by all levels of government, and estimates future spending needs to maintain or improve current conditions and performance.[5] As with any attempt to forecast future conditions, there is a host of simplifying assumptions, omissions, and data problems that influence the estimates of future funding needs. Among other things, the estimates of future needs rely on forecasts of travel demands and assume that the most economically productive projects (i.e., projects with the highest benefits relative to costs) will be implemented first. Despite such uncertainties and assumptions, these estimates provide a way to assess the level of current spending compared with what would be needed in the future under different scenarios.

The 2010 needs assessment, the most recent available, shows that, in 2008, $91.1 billion was spent on capital improvements to the nation's highways and bridges.[6] Of that amount, $76.8 billion was spent on roadways and $14.3 billion was spent on bridges. The vast majority of the expenditure on bridges, $12.8 billion, went to rehabilitate or replace existing bridges, with the remainder devoted to construction of new bridges.[7]

Because of the modeling involved, FHWA's future needs estimates for bridges are limited to fixing deficiencies in existing bridges only when the benefits outweigh the costs. The future needs estimate can therefore be measured against the $12.8 billion expenditure in 2008. The U.S. Department of Transportation (DOT) estimates that fixing all existing bridge deficiencies would cost $121.2 billion (in 2008 dollars).[8]

Of course, fixing all deficient bridges overnight is not feasible. FHWA, therefore, estimates how this investment backlog will change at various levels of spending over the 2008-2028 period, taking into account the deterioration of existing bridges over that period. The results of this analysis can be seen in *Table 1*. To keep the backlog at the 2008 level through 2028 would require $11.9 billion annually (in 2008 dollars), less than the level of spending in 2008. To eliminate the backlog by 2028, the maximum economically justified level of investment would be $20.5 billion annually, implying roughly a 4% annual increase in inflation-adjusted spending. Spending between $11.9 billion and $20.5 billion per year, FHWA estimated, would improve the conditions of the nation's bridges but would not entirely eliminate the investment backlog. At the level of spending in 2008, $12.8 billion per year, the total dollar cost of correcting all remaining deficiencies would decline by 11% by 2028.

Table 1. Projected Changes in 2028 Bridge Investment Backlog Compared with 2008 Levels for Different Possible Funding Levels

Annual Percentage Change in Spending	Average Annual Spending (Billion 2008 Dollars)	2028 Backlog (Billion 2008 Dollars)	Percentage Change from 2008	Funding Level Description
4.31%	20.5	0	-100.0%	Maximum economic investment scenario
3.51%	18.7	25.3	-79.1%	
2.88%	17.5	42.0	-65.3%	
1.31%	14.7	79.1	-34.7%	
0.56%	13.6	95.8	-20.9%	
0.00%	12.8	107.6	-11.2%	2008 spending on existing bridges
-0.70%	11.9	121.2	0%	Maintain investment backlog
-1.00%	11.5	127.1	4.9%	

Source: U.S. Department of Transportation, Federal Highway Administration and Federal Transit Administration, 2010 Status of the Nation's Highways, Bridges, and Transit: Conditions and Performance, exhibit 7-17.

FEDERAL AND STATE ROLES

Federal assistance for the maintenance, rehabilitation, and construction of highway bridges comes principally through the federal-aid highway program administered by FHWA. FHWA, however, does not make the determination as to which bridges should benefit from federal funding. Almost all funding under the federal-aid highway program is distributed to state departments of transportation, which determine, for the most part, where and on what the money is spent. States must comply with detailed federal planning guidelines as part of the decision-making process, but otherwise are free to spend their federal highway funds in any way consistent with federal laws and regulations. Bridge projects are developed at the state level, and state departments of transportation let the contracts, oversee the construction process, and provide for the inspection of bridges.[9]

The 2012 surface transportation reauthorization, the Moving Ahead for Progress in the 21st Century Act (MAP-21, P.L. 112-141), further strengthened the states' ability to determine spending on bridges by eliminating the

Highway Bridge Program (HBP), which provided money to the states specifically for bridge construction and rehabilitation. Bridge improvements remain eligible for funding under two programs created by MAP-21 that distribute funds to the states under formulas specified in the law, the National Highway Performance Program (NHPP) and the Surface Transportation Program (STP). Under both programs, the states determine how much of their federal funding goes for bridges as opposed to other uses, primarily highway construction and improvement. These funds may also be used for the seismic retrofitting of bridges to reduce earthquake failure risk.[10]

FHWA is involved in the project decision-making process in two significant ways. First, MAP-21 (§1111) requires FHWA, in consultation with the states and federal agencies, to classify public road bridges according to "serviceability, safety, and essentiality for public use ... [and] based on that classification, assign each a risk-based priority for systematic preventative maintenance, replacement or rehabilitation." However, none of the MAP-21 programs appear to require the new classification and risk-based priority metric be used to determine program eligibility.[11] In addition to developing this metric, FHWA imposes certain performance measures that states must meet to avoid funding penalties pursuant to MAP-21. For example, if more than 10% of the deck area of a state's bridges on the National Highway System is structurally deficient, the state is subject to a penalty requiring it to dedicate an amount of its NHPP funds equal to 50% FY2009 HBP spending to bridge projects.[12]

While the HBP existed, bridge program apportionments—the money states were entitled to receive each year under the HBP—were trending upward. The obligation of funds for bridge projects, however, tended to be substantially lower than the apportioned amounts. The transfer by the states of HBP funding to other highway programs, while permitted by law, was controversial following the collapse of the I-35W Bridge in Minnesota in 2007. At the time, critics saw the widening gap between annual apportionments and obligations as evidence of state transfer of resources to nonbridge uses. However, bridge spending was an eligible expense under all the core formula programs, not just HBP. The totals obligated from all Federal-Aid Highway programs (FAHP) for bridge work exceeded the HBP gross apportionments for the years FY2007-FY2012. *Table 2* shows the gap between the gross apportionments and the amounts obligated under the HBP, as well as the total obligations from all FAHP sources for FY2007 through FY2012. Past spending on bridge improvement may provide useful benchmarks of state spending efforts during congressional oversight.

Table 2. HBP Apportionments/Obligations and Obligations from All FAHP Sources: FY2007-2012 (Dollars in Millions)

	FY2007	FY2008	FY2009	FY2010	FY2011	FY2012
HBP Apportionments (gross)	$5,041	$5,058	$5,177	$5,612	$5,897	$5,509
HBP Obligations	$3,761	$4,066	$4,212	$4,284	$4,193	$3,575
Total: All Federal-Aid Highway	$6,418	$6,837	$9,386	$8,472	$7,043	$6,014
Programs' Bridge Obligations						

Source: FHWA. FY2009-FY2011 total obligations reflect obligation of stimulus funds under the American Recovery and Reinvestment Act of 2009 (P.L. 111-5).

Note: For a detailed table of bridge obligations for these years, see Appendix B.

Under MAP-21, which eliminated the HBP, the states have even more freedom to decide how much of their federal surface transportation grants to spend on bridges. Although there is no freestanding bridge program under MAP-21, FHWA is able to track the obligation of federal funds for bridge activities using bridge improvement type codes in its FMIS database.

Bridge Inspection

Under the National Bridge Inspection Program (NBIP), all bridges on public roads longer than 20 feet must be inspected by state inspectors or certified inspection contractors, based on federally defined requirements. Federal agencies are subject to the same requirements for federally owned bridges, such as those on federal lands. Data from these inspections are reported to FHWA, which uses them to compile a list of deficient or functionally obsolete bridges. States may use this information to identify which bridges need replacement or repair.[13]

FHWA sets the standards for bridge inspection through the National Bridge Inspection Standards (NBIS).[14] The NBIS set forth how, with what frequency, and by whom bridge inspection is to be completed. The standards provide the following:

- Each state is responsible for the inspection of all public highway bridges within the state except for those owned by the federal government or Indian tribes. Although the state may delegate some bridge inspection responsibilities to smaller units of government, the responsibility for having the inspections done in conformance with federal requirements remains with the state.

- Inspections can be done by state employees or by certified inspectors employed by consultants under contract to a state department of transportation.
- Inspection of a federally owned bridge is the responsibility of the federal agency that owns the bridge.
- The NBIS set forth the standards for the qualification and training of bridge inspection personnel.
- In general, the required frequency of inspection is every 24 months. States are to identify bridges that require less than a 24-month frequency. States can also, however, request FHWA approval to inspect certain bridges on an up to 48-month frequency. Frequency of underwater inspection is generally 60 months but may be increased to 72 months with the FHWA permission.
- The most common on-site inspection is a visual inspection by trained inspectors, one of whom must meet the additional training requirements of a team leader. Damage and special inspections do not require the presence of a team leader.
- Load rating of a bridge must be under the responsibility of a registered professional engineer. Structures that cannot carry maximum legal loads for the roadway must be posted.

The vast majority of inspections are done by state employees or consultants working for the states. FHWA inspectors do, at times, conduct audit inspections to assure that states are complying with the bridge inspection requirements. FHWA also provides on-site engineering expertise in the examination of the reasons for a catastrophic bridge failure. However, FHWA bridge engineers have only limited time available for audits and other bridge oversight.

FHWA's Emergency Relief Program

The Emergency Relief Program (ER) provides funding for bridges damaged in natural disasters or that are subject to catastrophic failures from an outside source.[15] The program provides funds for emergency repairs immediately after the failure to restore essential traffic, as well as for longer-term permanent repairs.

ER is authorized at $100 million per year, nationwide. Funding beyond this is commonly provided for in supplemental appropriations acts. In the case of most large disasters, additional ER funds are provided in an appropriations bill, usually a supplemental appropriations bill.

The federal share for emergency repairs to restore essential travel during the first 180 days following a disaster is 100%. Later repairs, as well as permanent repairs such as reconstruction or replacement of a collapsed bridge, are reimbursed at the same federal share that would normally apply to the federal-aid highway facility. Recently, Congress has sometimes legislatively raised the federal share under the ER program to 100% (as happened with the I-35W collapse in Minnesota). As is true with other FHWA programs, the ER program is administered through state departments of transportation in close coordination with FHWA's division office in each state. ER was the source of funds for replacement of the I-5 Skagit River Bridge in Washington State that collapsed on May 23, 2013, after being struck by a truck that was hauling an oversized load.

ISSUES FOR CONGRESS

The I-5 bridge collapse on May 23, 2013, led to warnings that the large number of structurally deficient bridges indicates an incipient crisis,[16] even though the I-5 bridge itself was not structurally deficient.[17] FHWA data do not substantiate this assertion. The numbers of bridges classified as structurally deficient or functionally obsolete have fallen consistently since 1990, and the proportion of all highway bridges falling into one or the other category is the lowest in decades.

The condition of roads has not experienced the same degree of improvement as the condition of bridges. This raises the policy question of what priority should go to bridge repairs as opposed to roadway repairs. In MAP-21, Congress implicitly addressed this issue by giving states greater flexibility to use federal funding for roads or for bridges, at their discretion. By doing this, Congress chose not to mandate bridge spending levels sufficient to reduce the number of deficient bridges by a certain date or eliminate deficient bridges altogether (described in *Table 1*). Instead, responsibility for determining the amount that should be spent on bridges each year was assigned to the states.

A related issue is one of terminology. The terms "structurally deficient" and "functionally obsolete" are not synonymous with "unsafe." An effort to eliminate all structurally deficient bridges quickly could lead to inefficient spending if a significant percentage of these bridges do not actually have major safety problems. Under MAP-21 FHWA is to develop performance measures in regard to bridges. The speed of their development and the effectiveness of implementation will be oversight issues for Congress.

Federal Pressure for State Bridge Spending

To encourage state spending on structurally deficient bridges, MAP-21 sets a penalty threshold under the NHPP: any state whose structurally deficient bridge deck area on the National Highway System within the state's borders exceeds 10% of its total National Highway System bridge deck area for three years in a row must devote NHPP funds equal to 50% of the state's FY2009 Highway Bridge program apportionment to improve bridge conditions during the following fiscal year and each year thereafter until the deck area of structurally deficient bridges falls to 10% or below. Even if a state were required to spend more of its federal highway funding on bridges (and therefore less on roadway projects) due to this penalty, its mandated spending on deficient bridges would be less than was required prior to the enactment of MAP-21.

MAP-21 expires at the end of FY2014. Given the lags in state reporting and the time required to complete major bridge projects, it may not be clear at that point whether the states' desire to spend their STP or NHPP funds on nonbridge projects is obstructing the declared national policy of reducing the number of deficient bridges. It is conceivable that Congress will begin debating surface transportation reauthorization with limited data on the success or failure of MAP-21 in regard to bridge improvement.

Providing More Money for Bridges

Federal motor fuel tax revenues, which have provided most of the funding for the federal-aid highway program since 1956, have been insufficient to support the program as authorized by Congress for several years. MAP-21 allocated money from the Treasury's general fund for highway and bridge programs in FY2013 and FY2014. If it wishes to increase spending on bridges following the expiration of MAP-21, Congress has a number of options:

- Provide general fund monies to accelerate the repair of the remaining structurally deficient and functionally obsolete bridges.
- Consider resurrecting a stand-alone program for structurally deficient bridges, which would essentially reverse the change made in MAP-21 and would force the states to provide minimum spending levels for bridge maintenance and repair.
- Raise the fuel taxes that finance the vast majority of surface transportation outlays, possibly with a portion of the increase dedicated just to a federal bridge program.
- Emphasize public-private partnerships (PPPs) as a mechanism to help reduce the number of structurally deficient bridges, for example, by allowing states to offer long-term leases of toll facilities to private investors in return for large up-front payments that could be used to supplement normal state and federal spending on bridge replacement and repair.

Encourage Tolling of Nontolled Bridges

Heavily traveled bridges can be attractive targets for conversion to toll facilities: many bridges have no convenient alternatives, so many drivers may be unable to avoid paying whatever toll is imposed. Congress might consider tolling as a means of accelerating the pace of bridge repair under the current constrained budgetary environment. An expansion of tolling could allow for more rapid improvement of major bridges. The revenue stream provided by tolls can also make bridge building and reconstruction an attractive investment for private entities that are interested in participating in a PPP. The revenue stream also can help projects become eligible for a federal Transportation Infrastructure Finance and Innovation Act (TIFIA) loan. Bridge tolls, however, are often very unpopular, and their acceptance varies greatly from region to region. Some states have sought to make bridge tolls more acceptable within a state by charging out-of-state users at a much higher rate than in-state residents, a practice that may face legal challenges.

Redirect Spending Away from Off-System Bridges

Historically, nearly all federal highway funding was restricted to roads and bridges on the federal-aid highway system. The Surface Transportation Assistance Act of 1978 (P.L. 95-599) stipulated that not less than 15% of a state's bridge apportionments nor more than 35% be spent "off-system." Off-system spending of federal bridge funds has been part of every highway authorization bill ever since. Under MAP-21, STP funds not less than 15% of the amounts apportioned to a state for the Highway Bridge Program in FY2009 are to be obligated for off-system bridge projects.

Off-system bridges, by definition, are inherently local in nature. By eliminating the set-aside for off-system bridges, Congress could enable states to spend more of their federal funds on bridges that are more heavily used, but states would not be required to spend funds for that purpose without additional legislation.

Maintenance

The FHWA requirement that federal funding used for bridges be directed to bridges with relatively low sufficiency ratings may encourage states to substitute bridge replacement for maintenance-type projects. During FY2011, of the total obligation of federal funds from all FHWA sources, 15% was obligated for new bridges, 56% was obligated for bridge replacement, 3% was for major rehabilitation, and 24% was for minor bridge work. Although these figures indicate that the lion's share of bridge funding has been obligated for new and replacement bridges, these percentages are less than they were in the late 1990s. The percentage spent on minor bridge work has increased significantly since then.[18] Still, the case can be made that as the number of deficient bridges decreases, rather than reducing bridge spending it might make sense to shift the focus on the spending over time toward preventive maintenance.

Oversight and Inspection Issues[19]

Risk-Based Approach to Federal Bridge Oversight

MAP-21 requires that the National Bridge Inventory classify bridges according to serviceability, safety, and essentiality for public use and, based on this classification, assign each bridge a risk-based priority for systematic preventative maintenance, replacement, or rehabilitation.

The risk-based approach would provide an additional metric to the traditional focus on bridges that are "structurally deficient" and "functionally obsolete." In particular, the risk-based approach, which is still under development by FHWA, could provide statistics that more clearly identify unsafe bridges. Once the metric is developed, Congress could consider making its use an eligibility requirement for bridge project funding under NHPP and STP.

Oversight of State Transportation Implementation Plans (STIPs)

MAP-21 maintains the previous requirement that states' spending of federal funds on bridges be based on priorities established in state transportation implementation plans (STIPs). Following the elimination of the Highway Bridge Program in 2012, Congress may want to examine state spending on bridges under MAP-21 and, in particular, whether STIPs pay adequate attention to bridge needs as opposed to highway needs.

Inspection Auditing

FHWA could be directed to take a more active role in ensuring that inspections done by the states or their contractors are done in conformance with the National Bridge Inspection Standards, including on-site audits of state inspections.

However, to have an impact, FHWA would have to be provided with sufficient funding to hire additional engineers and support personnel at FHWA Division offices and dedicate these resources to oversight of the inspection program.

Inspector Training and Personnel Qualifications

MAP-21 included requirements for establishment of minimum inspection standards and an annual review of state compliance with the standards established in the act. Within two years of enactment the Secretary of Transportation is to update the standards for the methodology, training, and qualifications of inspectors. Congress may wish to oversee implementation of these provisions.

APPENDIX A. BRIDGE CONDITION BY STATE

Table A-1. Bridge Condition by State as of December 2012

State	All Bridges (number)	Structurally Deficient (number)	Functionally Obsolete (number)	Percent of Bridges in State	
				Structurally Deficient	Functionally Obsolete
Alabama	16,070	1,448	2,205	9%	14%
Alaska	1,173	128	147	11%	13%
Arizona	7,835	247	721	3%	9%
Arkansas	12,696	898	2,031	7%	16%
California	24,812	2,978	4,178	12%	17%
Colorado	8,591	566	907	7%	11%
Connecticut	4,208	406	1,070	10%	25%
Delaware	862	53	122	6%	14%
District of Columbia	239	30	155	13%	65%
Florida	11,982	262	1,764	2%	15%
Georgia	14,739	878	1,871	6%	13%
Hawaii	1,131	146	359	13%	32%
Idaho	4,214	397	440	9%	10%
Illinois	26,514	2,311	1,976	9%	7%
Indiana	18,789	2,036	2,188	11%	12%
Iowa	24,496	5,193	1,282	21%	5%
Kansas	25,176	2,658	1,959	11%	8%
Kentucky	14,031	1,244	3,219	9%	23%
Louisiana	13,175	1,783	2,032	14%	15%
Maine	2,408	356	436	15%	18%
Maryland	5,294	368	1,099	7%	21%
Massachusetts	5,120	493	2,214	10%	43%
Michigan	11,000	1,354	1,672	12%	15%
Minnesota	13,121	1,190	423	9%	3%
Mississippi	17,061	2,417	1,357	14%	8%
Missouri	24,334	3,528	3,365	14%	14%
Montana	5,120	399	509	8%	10%
Nebraska	15,393	2,779	1,058	18%	7%

Table A-1. (Continued)

State	All Bridges (number)	Structurally Deficient (number)	Functionally Obsolete (number)	Percent of Bridges in State	
				Structurally Deficient	Functionally Obsolete
Nevada	1,798	40	216	2%	12%
New Hampshire	2,429	362	445	15%	18%
New Jersey	6,554	651	1,717	10%	26%
New Mexico	3,924	307	350	8%	9%
New York	17,420	2,169	4,718	12%	27%
North Carolina	18,165	2,192	3,296	12%	18%
North Dakota	4,453	746	247	17%	6%
Ohio	27,045	2,462	4,311	9%	16%
Oklahoma	23,781	5,382	1,604	23%	7%
Oregon	7,633	433	1,341	6%	18%
Pennsylvania	22,669	5,540	4,370	24%	19%
Rhode Island	757	156	255	21%	34%
South Carolina	9,271	1,141	840	12%	9%
South Dakota	5,870	1,208	237	21%	4%
Tennessee	19,985	1,195	2,669	6%	13%
Texas	52,260	1,372	8,680	3%	17%
Utah	2,947	126	343	4%	12%
Vermont	2,727	288	643	11%	24%
Virginia	13,769	1,250	2,421	9%	18%
Washington	7,840	366	1,693	5%	22%
West Virginia	7,093	952	1,595	13%	22%
Wisconsin	14,057	1,157	779	8%	6%
Wyoming	3,101	426	287	14%	9%
Puerto Rico	2,248	282	932	13%	41%
Total (incl. Puerto Rico)	607,380	66,749	84,748	11%	14%

Source: U.S. Department of Transportation, Federal Highway Administration, National Bridge Inventory, Deficient Bridges by State and Highway System, Washington, DC.

APPENDIX B. BRIDGE OBLIGATIONS BY PROGRAM: FY2007-FY2012

Table B-1. Bridge Obligations by Program FY2007-FY2012

Program	FY2007	FY2008	FY2009	FY2010	FY2011	FY2012	Total, FY2007-FY2012
Interstate Maintenance	566,367,740	531,148,044	456,257,769	659,096,900	583,304,527	755,656,556	3,020,683,491
National Highway System	629,914,133	870,072,229	597,997,506	863,300,679	836,649,803	680,253,396	3,608,115,518
Surface Transportation Program	476,908,635	547,815,377	708,246,051	603,721,498	586,685,394	558,073,243	2,933,634,820
Bridge Programs	3,761,052,215	4,066,121,536	4,211,724,679	4,283,730,495	4,193,314,245	3,575,482,507	20,025,304,142
Congestion Mitigation And Air Quality	23,905,076	52,369,318	8,579,895	47,636,428	91,470,609	(10,213,853)	161,378,157
Appalachian Development Highway System	19,971,806	449,969	61,133,266	30,653,664	28,236,759	5,436,959	145,432,455
Recreational Trails	—	—	—	—	—	—	—
Metropolitan Planning	—	—	—	—	—	—	—
1% Metropolitan Planning	—	—	—	—	—	—	—
High Priority Projects	141,223,886	188,500,355	226,877,040	150,934,801	224,452,978	61,045,589	804,534,295
Minimum Guarantee—TEA-21	70,261,361	(6,841,861)	(5,295,640)	(14,994,995)	(16,498,678)	12,053,469	45,525,517
Equity Bonus Exempt Lim	55,196,232	23,363,153	96,050,658	35,326,437	14,007,551	59,268,059	259,848,937
Coordinated Border Infrastructure Program	41,711	11,580,237	23,208,473	23,039,215	30,457,277	10,461,126	87,207,802
Safe Routes To School	—	—	—	—	—	—	—

Table B-1. (Continued)

Program	FY2007	FY2008	FY2009	FY2010	FY2011	FY2012	Total, FY2007-FY2012
Planning And Research	—	—	—	—	—	(200,000)	(200,000)
All Others	673,252,684	552,598,820	3,000,825,716	1,789,136,040	470,519,916	306,635,541	6,240,369,898
Total	6,418,095,480	6,837,177,177	9,385,605,414	8,471,581,163	7,042,600,382	6,013,952,592	37,331,835,031

Source: Federal Highway Administration.

End Notes

[1] The standards, authorized at 23 U.S.C. §144, cover bridges located on public roads that are 20 feet (6.1 meters) in length or longer. U.S. Department of Transportation, Federal Highway Administration, "Bridges by Owner, December 2012," National Bridge Inventory, http://www.fhwa.dot.gov/bridge/britab.cfm.

[2] Arterials, including Interstates, are roads designed to provide for relatively long trips at high speed and usually have multiple lanes and limited access. U.S. Department of Transportation, Federal Highway Administration, "Functional Classification of Bridges by Highway System, 2012 Count," National Bridge Inventory, http://www.fhwa.dot.gov/ bridge/britab.cfm.

[3] U.S. Department of Transportation, Federal Highway Administration, "Functional Classification of Bridges by Highway System, 2012 ADT," National Bridge Inventory, http://www.fhwa.dot.gov/bridge/britab.cfm.

[4] U.S. Department of Transportation, Federal Highway Administration, and Federal Transit Administration, 2010 Status of the Nation's Highways, Bridges, and Transit: Conditions and Performance, Washington, DC, 3-10, http://www.fhwa.dot.gov/ policy/2010 cpr/pdfs.htm.

[5] The "improve" scenario is the level of spending in which the investment is made in all projects for which the economic benefits are equal to or greater than the economic costs.

[6] These spending figures do not include routine maintenance costs.

[7] U.S. Department of Transportation, Conditions and Performance, 2010, exhibit 6-11.

[8] U.S. Department of Transportation, Conditions and Performance, 2010, 7-27.

[9] See CRS Report R42793, Federal-Aid Highway Program (FAHP): In Brief, by Robert S. Kirk.

[10] See CRS Report R41746, Earthquake Risk and U.S. Highway Infrastructure: Frequently Asked Questions, by William J. Mallett, Nicole T. Carter, and Peter Folger.

[11] Leftover funding from the Highway Bridge Program will continue to use the "sufficiency rating" for prioritizing project eligibility. For more information see http://www.fhwa. dot.gov/bridge/bridgeload01.cfm. See also definitions of structurally deficient and functionally obsolete at http://www.fhwa.dot.gov/bridge/0650dsup.cfm.

[12] For a definition of the National Highway System see http://www.fhwa.dot.gov/planning/ national_highway_system/.

[13] The National Bridge Inspection Program was initiated in 1968 following the 1967 collapse of the so-called Silver Bridge over the Ohio River. The National Bridge Inspection Standards were first issued in 1971. See, Federal Highway Administration, Tables of Frequently Requested NBI Information, http://www.fhwa.dot.gov/bridge/britab.cfm.

[14] 23 C.F.R. 650 subpart C.

[15] For a more detailed discussion of the ER program, see CRS Report R42804, Emergency Relief Program: Federal-Aid Highway Assistance for Disaster-Damaged Roads and Bridges, by Robert S. Kirk.

[16] See, for example, the Associated Press article "Many U.S. bridges at risk of failure like Interstate 5 collapse," Plain Dealer, May 26, 2013, http://www.cleveland.com/nation/ index.ssf/2013/05/many_us_bridges_at_risk_of_fai.html; "Washington bridge collapse serves as a wake-up call," USA Today, May 28, 2013, "Bridge collapse shines light on aging infrastructure," USA Today, May 24, 2013, http://www.usatoday.com/story/news/ nation/2013/05/24/washingtonbridge-collapse-nations-bridges-deficient/2358419/; Bryce Covert, "Washington Bridge Collapse Another Sign That America's Infrastructure Is In Bad Shape," Think Progress, http://thinkprogress.org/economy/2013/05/24/2058241/ seattle-

bridge-collapse-infrastructure/?mobile=nc; Angela Greiling Keane and James Nash, "I-5 Bridge Collapse Shows Bridge Repair Needs Across U.S.," Bloomberg, May 25, 2013, http://www.bloomberg.com/news/2013-05-24/ bridge-collapse-accents-structural-decay-as-budgets-sag.html.

[17] See National Bridge Inventory: Structure Inventory and Appraisal; WA Structure: 00004794A000000, Federal Highway Administration.

[18] Federal Highway Administration, "Obligation of Federal Funds for Bridge Projects Underway by Improvement Type," Highway Statistics, Washington, FHWA, various years, and Highway Statistics 2011, Table FA-10.

[19] See also Federal Highway Administration, Tables of Frequently Requested NBI Information, http://www.fhwa.dot.gov/bridge/britab.cfm.

In: U.S. Bridge Conditions …
Editor: Ruby L. Lacey

ISBN: 978-1-63117-486-5

Chapter 2

LTBP BRIDGE PERFORMANCE PRIMER*

John Hooks*[†] *and Dan M. Frangopol*[‡]*, Sc.D. h.c.
(JM Hooks & Associates)
(Lehigh University)

FOREWORD

This study was conducted as part of the Federal Highway Administration's Long-Term Bridge Performance (LTBP) Program. The LTBP Program is a minimum 20-year research effort to collect scientific performance field data, from a representative sample of bridges nationwide, that will help the bridge community better understand bridge deterioration and performance. The products from this program will be a collection of data-driven tools, including predictive and forecasting models that will enhance the abilities of bridge owners to optimize their management of bridges.

This report is intended to provide a comprehensive definition of bridge performance that will be the foundation for carefully designed research studies in the LTBP Program. The report describes the barriers and complications that hinder the understanding of bridge performance and identifies the measures by which bridge performance is currently defined. The report divides bridge performance into specific issues, identifies the most critical issues, and

* This is an edited, reformatted and augmented version of report no. FHWA-HRT-13-051, issued by the Federal Highway Administration, Office of Infrastructure Research and Development, dated December 2013.

describes the types of data necessary to analyze these issues. This report will be of interest to engineers involved with research, design, construction, inspection, maintenance, and management of bridges as well as to decisionmakers at all levels of management in public highway agencies.

Jorge Pagán-Ortiz
Director, Office of Infrastructure
Research and Development

ABSTRACT

The performance of bridges is critical to the overall performance of the highway transportation system in the United States. However, many critical aspects of bridge performance are not well understood. The reasons for this include the extreme diversity of the bridge infrastructure, the widely varying conditions under which bridges serve, and the lack of reliable data needed to understand performance. The Long-Term Bridge Performance (LTBP) Program was created to identify, collect, and analyze research-quality data on the most critical aspects of bridge performance. This report describes the bridge infrastructure in the United States and explains why reliable indicators of bridge performance are necessary. Current methods of measuring performance are described, and the need for more precisely targeted performance indicators is discussed. The report explains bridge performance in terms of the aspects that have the most impact on the ability of the bridge to serve its intended purpose in a safe, efficient, and economical manner. The purpose of this report is to provide an understanding of what bridge performance means and the factors that influence bridge performance. It describes the various ways in which bridge owners measure and report bridge performance. It also describes how the LTBP Program will create new measures to allow deeper analysis and understanding, which will lead to ways to improve long-term bridge performance.

LIST OF ACRONYMS

AASHTO	American Association of State Highway and Transportation Officials
ADT	Average daily traffic
ASD	Allowable Stress Design

BCI	Bridge condition index
FHWA	Federal Highway Administration
Finnra	Finnish National Road Administration
FO	Functionally Obsolete
LCV	Lack of capital value
LRFD	Load and Resistance Factor Design
LTBP	Long-Term Bridge Performance
NBIS	National Bridge Inspection Standards
NBI	National Bridge Inventory
NHS	National Highway System
SD	Structurally Deficient
SHM	Structural health monitoring
SR	Sufficiency Rating
STRAHNET	Strategic Highway Network

SECTION 1. THE BASICS OF BRIDGE PERFORMANCE

The United States is more dependent than ever on its transportation system to consistently support a thriving economy and to afford its citizens a satisfactory quality of life. The system must provide for rapid response in case of personal or public emergencies; ensure a high level of national security, safety, and resilience; and provide the ability to respond quickly in times of attack or natural disaster. However, the engineered components of the transportation infrastructure, including bridges, have aged. Some are deteriorated, and many have exhausted their capacity to meet the ever-expanding operational demands in urban areas and along major freight routes. The system has also become more interdependent on other transportation modes and on financial, telecommunication, and cyber infrastructures for its operational and structural safety and security.

The transportation user community expects and deserves a system that routinely provides the highest quality service in terms of safety, efficiency, and economy while having the least possible impact on the local and global environment. Optimal operation of public highway systems is dependent on many factors. Bridges are critical nodes in the highway infrastructure, and poor performance of any bridge has the potential to reduce the operating capacity of the highway system it is a part of. Under current circumstances, most bridges will eventually reach a state where work is necessary to return the bridge or some of its components to a satisfactory level of condition or safety. The true criticality of bridges is often only apparent when work is necessary to maintain, rehabilitate, or replace an existing bridge or series of bridges; a bridge is closed due to conditions or service below acceptable levels; or a bridge collapses with attendant damage to property, disruption to service, or loss of life. Work zones where bridge work is underway usually involve one or more conditions that result in disruption to safe, efficient, and economical traffic flow. These conditions include narrowed or closed lanes, live load restrictions, speed reductions, inefficient detours, and safety hazards that result from these conditions. Negative impacts on local and regional economies and environments can often result from loss of productive time due to traffic delays and detours, work-zone accidents, or increased consumption of fuel and engine emissions, among other issues.

In the United States, the current knowledge base of inventory and condition data on bridges is among the best in the world. Current programs and methods of bridge inspection and the tools used for managing bridge programs are also among the best in the world. Yet the level of understanding of how bridges perform and how to satisfactorily measure their performance falls well short of desirable. Many attempts at performance assessment rely on expert opinion combined with significant assumptions and generalizations.

This bridge performance primer is intended to provide a clear and comprehensive perspective on the concept of bridge performance and to present a path to a better and deeper understanding of bridge performance, including how it can be reliably measured and used to improve the highway transportation system. This report describes the various ways in which bridge owners can and do currently measure and report bridge performance. It also describes how the Long-Term Bridge Performance (LTBP) Program will create new measures of performance that will allow a deeper analysis and understanding of performance, which will lead to ways to improve long-term bridge performance.

The Highway Bridge Infrastructure

The highway bridge infrastructure in the United States is very large and diverse. In this report, the term *bridge* is defined by the National Bridge Inspection Standards (NBIS): “a structure including supports erected over a depression or obstruction such as water, highway, or railway and having a track or passageway for carrying traffic or other moving loads and having an opening measured along the center of the roadway of more than 20 ft.”[1]

A brief look at data from the National Bridge Inventory (NBI) provides a revealing picture of this vital public asset.

The 2011 NBI contains records for 605,098 bridges, of which 132,150 are classified as tunnels or culverts. [2] The remaining 472,948 are single- or multi-span bridges separating vehicular traffic from other traffic or some topographical feature, usually a stream or river. These bridges range from the average highway overpass structure to "signature bridges" such as the Golden Gate Bridge, the Brooklyn Bridge, or the Sunshine Skyway. Within this range, the diversity of the bridge infrastructure in terms of age and design parameters (including structural type, materials of construction, width, and length) is exceptionally broad. NBI records describe bridges using many different attributes or parameters. Table 1 provides an abbreviated list of these characteristics and indicates how many different types there are in the NBI for each. [1]

Table 1. Diversity of Bridge Characteristics [1]

NBI Item	Number of Types
Kind of material, main span and/or approach span	10
Structure type, main span and/or approach span	23
Design load	10
Bridge posting	6
Deck structure type	9
Wearing surface	9
Membrane	5
Protective system	9

In addition to these mostly fixed attributes or parameters, there are wide variations in the ages of bridges and the service conditions under which they perform. Figure 1 illustrates the diversity in bridge ages by providing a histogram of bridges still in service grouped in 5-year increments. The average age of all bridges in the NBI is 41.0 years. One of the oldest bridges in the NBI is in Sackets Harbor, NY, carrying Military Road over Mill Creek. It is a concrete arch bridge with a deck that is 44.9 ft long and 21.6 ft wide. The bridge was built in 1800 and underwent a complete rehabilitation in 2003. As of 2011, the bridge was rated in excellent condition and was open to traffic with no load restrictions. [2]

The National Highway System (NHS) of the United States comprises approximately 160,000 mi of roadway, including the Interstate Highway

System and other roads, which are important to the nation's economy, defense, and mobility. The total mileage of the NHS includes only 4 percent of the nation's roads, but it carries more than 40 percent of all highway traffic, 75 percent of heavy truck traffic, and 90 percent of tourist traffic. About 90 percent of the U.S. population lives within 5 mi of an NHS road. All urban areas with a population of more than 50,000 and 93 percent of areas with a population of between 5,000 and 50,000 are within 5 mi of an NHS road. Counties that contain NHS highways also host 99 percent of all jobs in the nation, including 99 percent of manufacturing jobs, 97 percent of mining jobs, and 93 percent of agricultural jobs. There are 116,523 bridges on the NHS. [2,3]

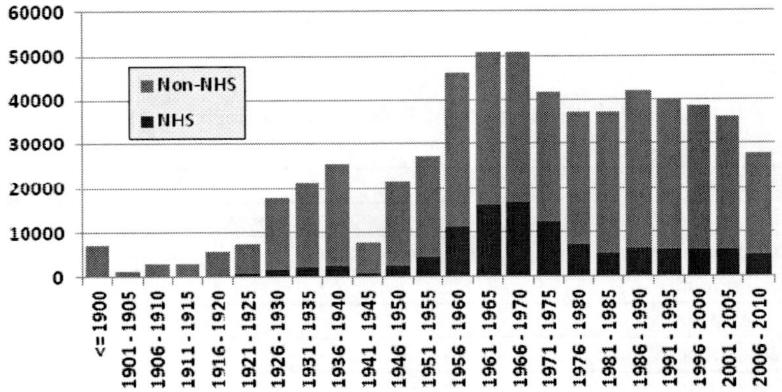

Figure 1. Graph. Age Distribution of All Bridges in the United States [2].

The age of a bridge serves as a rudimentary measure of the aggregate service provided during its service life as well as a harbinger of the level of degradation and damage the bridge may have experienced without consideration of periodic interventions for preservation, maintenance, repair, or major reconstruction. Age data can also be meaningful in terms of understanding the bridge materials, design standards and specifications, and construction processes used in construction of a bridge. The overall performance of bridges of a certain age might be expected to be poorer than that of bridges built after more recent advancements. Advancements generally correlate with better current and future performance, though this is never a certainty.

Service conditions for bridges vary dramatically in terms of traffic volume; truck loading (including overweight permit loads); level of vulnerability to natural events such as floods, ice, impact of waterborne debris, wind, and seismic loading; and susceptibility to long-term effects of climate and the service environment. Often, there is also the potential for significant damage from manmade hazards such as impacts by waterborne vessels or vehicular traffic, fires, explosions, chemical spills, or other events not considered ordinary service factors. In regions where snow and ice control is necessary, routine winter deicing operations result in application of chemicals that are known agents of deterioration of bridge elements.

Other critical factors affecting bridge performance are the types, frequency, and effectiveness of preservation, maintenance, repair, or rehabilitation actions performed on bridges by the owner. In a few States, the transportation department owns and is responsible for maintenance of all bridges on public roads. However, in most States, a significant percentage of the bridges on public highways are owned by other types of agencies, including agencies at the county and city levels of government, railroad companies, toll authorities, and other semiprivate or private entities. Table 2 shows the numbers of different types of entities that have maintenance responsibilities for bridges on public highways. Bridge performance is impacted significantly by the policies and procedures at any bridge-owning agency as well as by the agency's culture, the knowledge and experience of bridge personnel, the priorities of the agency, the revenues available for bridge programs, and the agency's funding mechanisms.

Table 2. Types of Entities with Bridge Maintenance Responsibility [1]

Entity Category	Number of Different Types
State and local highway agencies	4
Other State and local agencies	4
Private owners	4
Federal agencies	15

The cumulative effect of these factors may vary. At the level of State transportation departments, performance is generally good to very good, whereas at the level of local agencies, where resources are limited, performance may be fair to poor. More detailed information on the demographics of the bridge infrastructure in the United States, as recorded in the 2011 NBI, can be found in the appendix.

The Federal Highway Administration (FHWA) has initiated the LTBP Program, which was authorized in the *Safe, Accountable, Flexible, and Efficient Transportation Equity Act: A Legacy for Users*, signed into law in August 2005. The LTBP Program is a minimum 20-year, multifaceted research effort that is strategic in nature and has both specific short-term and long-range goals. Its concept is similar to the Long-Term Pavement Performance program that has been underway for more than 20 years. The overall objective of the LTBP Program is to inspect, evaluate, and periodically monitor representative samples of bridges nationwide in order to collect, document, maintain, and manage high-quality, quantitative performance data over an extended period of time. The program will employ sensing technologies and non-destructive evaluation and testing tools in addition to typical bridge inspection approaches. It will also require close collaboration among stakeholders, including public agencies, academia, and industry. Data collected under the LTBP Program will help in the performance evaluation of certain critical elements or features of bridges such as decks. Tools such as the Bridge Portal, data-driven deterioration models, life-cycle cost models, and a bridge condition index (BCI) will be developed to help engineers analyze the long-term performance of individual bridges.

The Meaning of "Bridge Performance"

One simple definition of *performance* is that performance equates to accomplishment of a specified purpose or set of purposes. For the purposes of the LTBP Program, the following definition is used: Bridge performance encompasses how bridges function and behave under the complex and interrelated factors they are subjected to day in and day out—traffic volumes, loads, deicing chemicals, freeze-thaw cycles, rains, or high winds. Bridge design, construction, materials, age, and maintenance history also play roles in performance. Performance is usually associated with some set of standards, whether absolute or relative, and performance can be measured against those standards. In baseball, for example, one purpose of a batter is to hit safely, and a measure of the batter's performance is the batting average—the number of hits divided by the number of at-bats. A batter hitting over 0.300 is considered to be performing very well.

The purpose of a bridge is inherent in the NBIS definition: to carry vehicular traffic or other moving loads over a depression or obstruction such as water, highway, or railroad.

Transportation users expect a given bridge to accomplish its purpose in a satisfactory manner as measured against several objectives. Bridges should do the following:

- Present a minimal safety hazard to users during normal service.
- Present minimal obstruction to the free flow of traffic at all times.
- Have a minimal negative impact on the local and global environments.
- Present an acceptable level of risk against catastrophic failure.
- Present an aesthetically pleasing appearance.
- Accomplish all of the preceding elements with minimal life-cycle costs.

As a performance indicator for a baseball batter, batting average is a simple index with a range of 0.000–1.000. The batting average is a useful measure, but it does not reflect the value of the batter to the baseball team in a complete manner. Other related factors are also important, including hits with runners in scoring position; number of bases reached via singles, doubles, triples, and home runs; percentage of times reaching base safely; or even the outcome of at-bats in which the batter does not hit safely. In the same manner, the overall performance of a bridge cannot be described by any one rating, index value, or other individual parameter.

The differences in batters' levels of performance are analogous to the differences in bridges' levels of performance. Imagine bridge A is located in an arid (dry, no-freeze) climate and bridge B is located in a northern (wet, freeze) climate. The bridges are steel multigirder simple span bridges that were built in the same year. The girders on bridge A are rated 8 on the NBI scale, nearly new condition. The girders on bridge B are rated 4, for poor condition with advanced section loss and deterioration. In comparing the relative performance of each bridge, factors such as annual snowfall, amount of deicing salts applied, and condition of the joints must be considered. Also of importance are the type and frequency of preservation treatments applied to the bridge (e.g., sealing joints, washing the superstructure, etc.). Continuing the analogy, baseball factors such as hits with runners in scoring position could be the equivalent of traffic and heavy trucks carried on the bridges.

There are many different ways to measure bridge performance. Some of these measures are in the form of an index value calculated from a defined formula using input data such as NBI condition ratings and traffic volumes. As detailed later in this report, numerical indices such as the Federal Sufficiency

Rating (SR), California Bridge Health Index, and Finnish National Road Administration (Finnra) Repair Index have been used for quite some time to define bridge performance. [2, 4, 5] For example, SR is an index calculated from an established formula and has a range of 100 (representing an entirely sufficient bridge) to 0 (representing an entirely insufficient or deficient bridge). An SR value of less than 50 indicates a low level of performance but does not reveal the specific characteristics of the bridge that render it deficient.

Other measures are based on whether or not a bridge meets some defined criteria. These include classification as Structurally Deficient (SD) or Functionally Obsolete (FO) and whether a bridge meets or exceeds defined deflection limits.

Still other measures reflect a level of operational capacity or service, including posted load, load rating, rideability of the bridge wearing surface, traffic congestion near or on the bridge, number of accidents on the bridge, and percentage of rain events and floods that overtop the bridge. Most of these measures can be applied to an individual bridge or to a population of bridges, such as bridges on the NHS or bridges of a certain type and material.

Section 2 and section 3 provide insight into some of these measures, including measures for special considerations related to bridge safety and stability. Each measure has value depending on the person using the measures and the purpose for using it. For example, commercial interests, shippers, and drivers can use simple measures of posted weight limits or geometrical dimension limitations to weigh performance in terms of durability, operational capacity, and roadway safety, all important factors in supporting a free flow of goods and services.

In contrast, section 4 shows that these measures do not readily support a better understanding of bridge performance because it is difficult to correlate changes in these measures with underlying causes of changes in performance. Section 4 discusses what needs to be done to better understand and develop solutions to improve bridge performance.

Experience has shown that the performance of any specific bridge is dependent on complex interactions of multiple factors, many of which are closely linked, including the following:

- The original design parameters and specifications such as bridge type, materials of construction, geometry, and load capacity.
- The initial quality of materials and of the as-built construction.
- The varying environmental conditions of climate, air quality, and surrounding soil.

- The extent and severity of corrosion or other deterioration processes that affect component strength or structural behavior (e.g., bearings locked up due to corrosion products).
- The traffic volumes and frequency and the weight of truck traffic carried by the structure.
- The type, timing, and effectiveness of preventive maintenance, minor and major rehabilitation actions, and ultimately, replacement actions.

All of these factors combine to affect the condition and operational capacities of the bridge and its structural elements at any given point in the life of the bridge. While simple measures such as those described in this section are currently used to evaluate overall performance, a better performance measure would be an indicator of the qualitative or quantitative impact of a parameter or set of parameters on some specific aspect of bridge performance.

Table 3. Main Categories of Bridge Performance and General Contributing Factors

Category	Important Data or Factor
Structural condition—durability and serviceability	Structure type
	Structural materials and material specifications
	Structure age
	As-built material qualities and current conditions
	As-built construction qualities and current conditions
	Truck loads and other live loads
	Environment—climate, air quality, and marine atmosphere
	Snow and ice removal operations
	Type, timing, and effectiveness of preventive maintenance
	Type, timing, and effectiveness of restorative maintenance and minor and major rehabilitation
	Flooding, hydraulic design, and scour mitigation measures
	Subsurface soil characteristics—settlement
Functionality—user safety and level of service	Structure geometry—clear deck width, skew, and approach roadway alignment
	Skid resistance and ride quality of riding surface
	Vertical clearances—over and under
	Traffic volumes and percentage of trucks
	Posted speed

Table 3. (Continued)

Category	Important Data or Factor
Structural integrity — safety and stability in all failure modes	Seismic performance
	Hurricane and flood resistance
	Collision impacts
	Blast impacts
	Fire resistance
	Structural redundancy and load redistribution
Costs to users and agency	Users: Accident costs
	Users: Detour and delay costs
	Agency: Initial construction costs
	Agency: Maintenance, repair, and rehabilitation costs
	Agency: Traffic maintenance costs

Because bridge performance is a complex issue, it is useful to organize the primary issues relating to bridge performance into general categories. As shown in figure 2, the primary issues in bridge performance can be divided into four general categories: structural condition (for durability and serviceability), functionality (safety and traffic capacity), structural integrity and risk (for safety and stability), and costs to the user and to the agency.

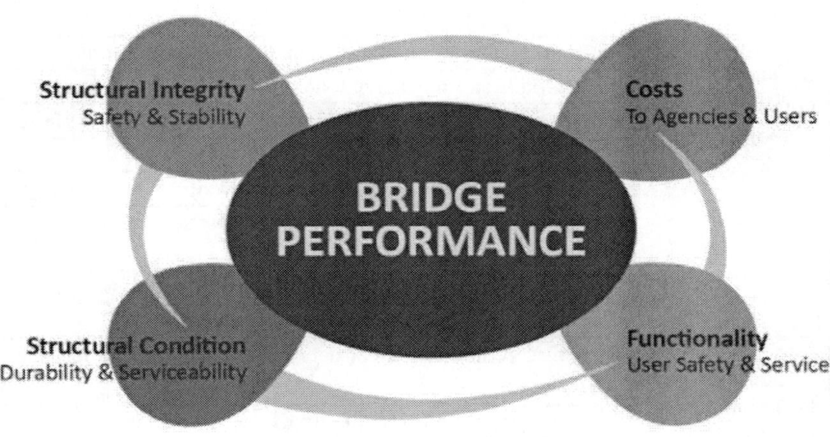

Figure 2. Illustration. Main Categories of Bridge Performance Issues.

Many relevant factors combine to affect performance under each of these four main categories. Table 3 lists the relevant factors that might impact various aspects of bridge performance. Within these categories are many specific performance issues that are of importance to the bridge community and that could be studied over the long term to achieve better understanding. For each of these specific performance issues, there are multiple data items that could be gathered to assist in the evaluation of performance.

Why Measure Bridge Performance?

Bridge performance is an issue that is of some concern to virtually everyone in the country, including people in the following roles:

- Highway users, including the following:
  - Commuters.
  - Routine travelers and tourists.
  - Deliverers of goods and services.
  - Emergency responders.
- Legislators who create programs and provide funds for the design, construction, inspection, maintenance, repair, and replacement of bridges.
- Policymakers and administrators in transportation agencies.
- Engineers and planners who design and build highways and bridges.
- Bridge maintenance and bridge management engineers and personnel who work to keep bridges at a satisfactory level of service and prevent failures.
- Consultants and commercial interests who provide materials and services for the design, construction, inspection, management, maintenance, repair, demolition, and replacement of bridges.
- Providers of news who inform the public of issues of local and national interest related to highways and bridges.

Bridge performance measures may have multiple uses depending on the perspective and responsibilities of those using the measures. The average everyday user seeks congestion-free travel, reassurance about highway safety, and rapid assistance from first responders.

Less apparent but also of great importance to the everyday user is minimization of the use of public resources to keep bridges safe and at a satisfactory level of service.

Engineers and planners should factor performance into future planning, design, and construction of bridges by applying lessons learned from the performance of previously built bridges. Bridge maintenance and bridge management personnel use measures of performance to evaluate the policies, practices, techniques, and materials that they employ to prolong bridge service at a satisfactory level and to project future bridge needs.

Consultants and commercial entities use performance indicators to make critical business decisions on what technology, equipment, materials, and services to develop or improve and to provide to bridge owners.

News outlets use simple performance measures and statistics to inform the general public and key transportation constituencies about critical issues related to the transportation system.

Engineers, planners, and bridge maintenance and management personnel have the most immediate use for reliable bridge performance measures because they can have the most immediate impact on changing the factors that influence bridge performance. These individuals and their organizations need reliable bridge performance measures in order to do the following:

- Accurately evaluate congestion and traffic safety.
- Provide an accurate measure of load capacity, safety, and the need for load restrictions.
- Identify clear links between specific policies, actions, and the resulting change in performance level of bridge features based on the following:
    - Improved knowledge of how and why bridges deteriorate (i.e., advances in deterioration and predictive models).
    - Better understanding of the effectiveness of various maintenance, repair, and rehabilitation strategies as well as management practices.
    - Better understanding of the effectiveness of durability strategies for new bridge construction, including material selection.
    - Improvements in bridge management practice using high-quality quantitative data.
- Evaluate serviceability and durability.
- Set priorities for resource allocations within the transportation system and the bridge infrastructure.

- Evaluate organization-wide policies and programs such as the split between maintenance and capital funds.
- Improve system reliability, redundancy, and accountability.
- Establish risk-based evaluations of bridges that are vulnerable to failure.

SECTION 2. CURRENT APPROACHES TO MEASURING BRIDGE PEFORMANCE

Making useful and reliable assessments of bridge performance is a challenging task. Many complex factors contribute to the difficulty of measuring bridge performance, including the following:

- The large scale and extreme diversity of the bridge infrastructure.
- The many and variable causative factors that impact performance.
- A limited understanding of some of the key cause-and-effect relationships in bridge deterioration or diminished functional capacity.
- Data that are inconsistent, not in a usable format, not easily accessible, or unavailable.
- Different policies and practices for design, construction, inspection, preservation, maintenance, repair, and replacement for different samples of bridges.
- Changes in design codes, construction practices, and traditional bridge materials over time.
- The introduction of new and innovative materials.
- The widely differing objectives of various groups who assess bridge performance.

The last point is an issue of considerable importance in the assessment of bridge performance. Bridge engineering is not a static art or science. As new materials and techniques are developed and implemented on bridges, the assessment of performance, particularly of structural condition and integrity, is altered.

Examples over the history of bridges include the following:

- Bolting and welding replaced riveting.
- Epoxy-coated reinforcing bars replaced black bars in many States.

- New alloys were developed to provide further corrosion protection for reinforcing bars.
- Design of concrete mixtures evolved, with significant improvements in strength and permeability characteristics.
- High-performance steels with greater strength, ductility, and corrosion resistance have become routinely used.
- High-performance concretes and steel have led to the possibility of lighter superstructure dead loads.
- Applications of non-traditional materials such as fiber-reinforced polymer composites have been developed both for new bridges and for repair or enhancement of in-service bridges.
- Repair materials and methods have changed and developed.
- Joint details have changed, including the use of integral abutment bridges.

Many aspects of performance are moving targets and present considerable difficulty in making reliable and consistent measurements. Regardless of the difficulties involved, developing and implementing reliable bridge performance measures is of paramount importance. The array of consumers of bridge performance knowledge is broad, and critical social, economic, commercial, and political decisions hinge on those measurements. Every State transportation department has developed a set of performance measures for its highway system. These performance measures address a wide variety of objectives, covering issues such as the following:

- Reductions in injuries, fatalities, and property damage.
- Improvements in mobility of people and goods.
- Reductions in incident-related delays.
- Increases in the percent of pavements and bridges in satisfactory or better condition.
- Increases in the percent of bridges that are not posted with a weight limit restricting use by legally loaded vehicles.
- Improved customer satisfaction with maintenance, ride quality, and incidents of congestion.

Many of these performance measures are dependent on the condition and functional operation of the owner's bridges. But evaluating the impact of bridges on the achievement of most of these objectives is difficult at best, requiring intuitive reasoning and significant assumptions about cause and effect.

Most States that have established overall agency and system performance measures have also established more specific measures related to bridges. Some transportation departments use measures related to the percentage of bridges at a certain defined level of condition, such as good or excellent. Often, these performance measures relate to numbers or deck areas of bridges that are deficient or involve some index calculated based on data in the NBI or on element-level inspection data collected for bridge management databases. Examples of measures in use include the following:

- The percentage of bridge structures on the State highway system having a condition rating of either excellent or good.
- The percentage of bridge structures on the State highway system with posted weight restrictions.
- The percentage of bridge structures on the State highway system with an SR over 80.

NBI Condition and Appraisal Ratings

For almost four decades, as required by NBIS, bridge owners in the United States have compiled a complete inventory of bridge information and condition data, and many State transportation departments have several years of experience compiling comprehensive bridge databases for use in their bridge management systems. Despite these multiyear efforts, the availability of high-quality, useful data on many of the factors impacting bridge performance varies significantly from State to State. Much of the fixed data (e.g., structure type, construction materials, dimensions, clearances, scour protection, functional classification) are well documented and easily accessible. Current and historical data on the physical condition of bridge elements are also readily accessible; however, these data have shortcomings that are discussed later in this report. Beyond these two types of data, the availability and accessibility of high-quality, useful data on factors impacting bridge performance are generally poor to fair.

In the United States, most of the effort in bridge performance assessment has concentrated on measuring, recording, analyzing, and using bridge condition data. Collection of detailed information and condition data about bridges in the United States started with implementation of NBIS in 1971 and has continued in a consistent, systematic, and computerized format for four decades. While these data have been and still are used to assess bridge performance, the NBI does not represent a complete basis for documenting and assessing long-term bridge performance with proper consideration of all relevant factors. A proper assessment of bridge performance over time requires systematic correlation of changes in bridge conditions and functional capacities with key policies, programs, and actions that affect those conditions and capacities.

The collapse of the Silver Bridge at Point Pleasant, WV, in December 1967 was the defining moment in the development of bridge inspection programs, bridge data collection, formal bridge improvement programs, and, ultimately, modern bridge management systems. Prior to this event, knowledge of bridges was poor. Immediately after the collapse, crucial questions about the bridge population arose: How many bridges are there in the United States? What types of structures? Of what materials were they constructed? Where are they located? What are their current conditions? How vulnerable are they to failure? What are the immediate improvement priorities? What is the required scale of effort and associated cost to address significant deficiencies? There were virtually no useful answers immediately available. As a consequence, there was no basis for assessing the condition of individual bridges or the overall bridge population. There was certainly no basis for assessing the performance of bridges over time.

The NBI was created to address the absence of knowledge about the bridge inventory and bridge conditions. However, the NBI was neither envisioned, nor structured, to support the assessment of bridge performance over the long term. Guidance on meeting the requirements of the NBI was published in 1971, and by the end of 1973, the States had inventoried most of the bridges on the Federal-Aid Highway systems. Over time, the NBI has become a current and historical database, comprising a consistent set of data on almost every bridge over 20 ft long on all public highways in the United States. As a comprehensive database of information about bridges spanning the majority of a continent, the NBI remains unique in the world. Inventory and condition data have been collected on a large population of individual bridges for almost four decades, and the guidelines for collecting, reporting, checking, editing, and storing these data have been consistent over the life of the NBI. The guidelines were carefully written, were accepted and used by all domestic transportation agencies, and have only been modified slightly when needed. In particular, the system of evaluating and recording the key appraisal and condition data on each bridge has remained virtually unchanged over the full life of the NBI. A full explanation of the data and the guidelines for collecting and recording data can be found in the current issue of *Recording and Coding Guide for the Structure Inventory and Appraisal of the Nation's Bridges*, published by FHWA. [1]

NBI data are categorized by type, as follows:

- Identification (location, features carried, and features crossed).
- Structure type and material.
- Age and time in service.
- Geometry.
- Navigation.
- Highway classification.
- Condition ratings.
- Load rating and posting.
- Appraisal ratings (current ratings of adequacy of major features such as deck width and approach roadway alignment).
- Proposed improvements and inspection requirements.

With regard to supporting performance assessment, the key NBI data are structure type and material, condition ratings, and appraisal ratings. Fields that are less important but still useful are traffic volume data, location information (indicative of climate and environmental factors and potential for corrosion), and load ratings. NBI condition and appraisal ratings are generally attributed according to a scale of 0–9.

In practice, ratings in the range of 4–7 are most common for bridges in service. As an example, applicable NBI codes and associated definitions for the condition rating of deck, superstructure, and substructure (NBI items 58–60) are as follows:

- **9, Excellent condition.**
- **8, Very good condition:** No problems noted.
- **7, Good condition:** Some minor problems.
- **6, Satisfactory condition:** Structural elements show some minor deterioration.
- **5, Fair condition:** All primary structural elements are sound but may have minor section loss, cracking, spalling, or scour.
- **4, Poor condition**: Advanced section loss, deterioration, spalling, or scour.

- **3, Serious condition:** Loss of section, deterioration, spalling, or scour have seriously affected primary structural components. Local failures are possible. Fatigue cracks in steel or shear cracks in concrete may be present.
- **2, Critical condition:** Advanced deterioration of primary structural elements. Fatigue cracks in steel or shear cracks in concrete may be present or scour may have removed substructure support. Unless closely monitored, it may be necessary to close the bridge until corrective action is taken.
- **1, Imminent failure condition:** Major deterioration or section loss present in critical structural components or obvious vertical or horizontal movement affecting structure stability. Bridge is closed to traffic but corrective action may put back in light service.
- **0, Failed condition:** Out of service—beyond corrective action.

Applicable NBI codes and definitions for the appraisal rating of the same items are as follows:

- **NA**: Not applicable.
- **9**: Superior to present desirable criteria.
- **8**: Equal to present desirable criteria.
- **7**: Better than present minimum criteria.
- **6**: Equal to present minimum criteria.
- **5**: Somewhat better than minimum adequacy to tolerate being left in place as is.
- **4**: Meets minimum tolerable limits to be left in place as is.
- **3**: Basically intolerable requiring high priority of corrective action.
- **2**: Basically intolerable requiring high priority of replacement.
- **1**: This value of rating code is not used.
- **0**: Bridge closed.

While these performance measures have served well when used for condition evaluation and apportionment of funds, these measures do not cover all of the variables that researchers would like to evaluate.

Performance Measures Based on Bridge Deficiencies

The NBI condition ratings are well established after almost four decades of use in assessing the current condition of the major components of bridges being inventoried and inspected. The same is true of NBI appraisal ratings for assessing functional capacities. Changes in these ratings over time reflect the general performance of the bridge. The ratings are used to classify bridges as deficient or not deficient.

Bridges with low NBI condition or appraisal ratings are flagged and classified as follows:

- **SD:** A highway bridge is classified as *structurally deficient* if item 58 (deck), item 59 (superstructure), item 60 (substructure), or item 62 (culvert) is rated "poor" condition or worse (coded 4 or lower on the NBI rating scale). A bridge can also be classified as SD if its load-carrying capacity is significantly below current design standards, with item 67 (structural evaluation appraisal) coded 2 or lower, or if item 71 (waterway adequacy) for the feature below the bridge is coded 2 or lower.
- **FO:** A highway bridge classified as *functionally obsolete* is not SD, but its design is outdated. The bridge may have lower load-carrying capacity, narrower shoulders, or less clearance underneath than bridges built to the current standards. Classification as FO is triggered by a code of 3 or lower for item 68 (deck geometry appraisal), item 69 (underclearances, vertical and horizontal), or item 72 (approach roadway appraisal). A bridge is also classified as FO if item 67 (structural evaluation appraisal) or item 71 (waterway adequacy appraisal) is coded 3.

The two classifications of deficient bridges provide simple tools for agencies to describe the overall performance of their bridge populations and the overall effectiveness of their bridge programs. An unpublished 1999 FHWA study described various ways in which State transportation departments use these two classifications in their measurement of bridge performance. Several agencies use only the number of deficient bridges as a measure, whereas others use a combination of SR and the number of deficient bridges. When determining the proportion of a bridge population flagged as deficient, some agencies take into consideration the relative size of deficient bridges, represented by the cumulative square footage of deck the structures carry, rather than a simple count of structures. This approach is intended to discourage the practice of repairing small structures to improve the metrics while deferring repairs on larger structures that carry significant traffic for which repair would incur greater expense. Oftentimes, agencies will consider both metrics in their assessment of program performance.

The Federal SR

SR is an index that was devised by FHWA and used to evaluate the eligibility of bridges for Federal highway bridge rehabilitation and replacement funds. The SR formula is a method of evaluating highway bridge data by calculating and summing four separate factors to obtain a numeric value indicative of bridge sufficiency to remain in service. On the resulting rating scale, 100 represents an entirely sufficient bridge and 0 represents an entirely insufficient or deficient bridge.

SR is calculated using a complex formula wherein weighting factors are assigned to several bridge parameters and attributes in order to arrive at the numerical index for each bridge. The basic formula is shown in figure 3.

$$SR = S_1 + S_2 + S_3 - S_4$$

Figure 3. Equation. Federal SR [1].

Where:

S_1 = Structural adequacy and safety (maximum value = 55 percent).

S_2 = Serviceability and functional obsolescence (maximum value = 30 percent).

S_3 = Essentiality for public use (maximum value = 15 percent).

S_4 = Special reductions (maximum value = 6 percent).

These four factors provide consideration and weight to the following:

- **Structural adequacy and safety**: Condition ratings for deck, superstructure, and substructure plus the inventory (load) rating.
- **Serviceability and functional obsolescence:** Traffic lanes, average daily traffic (ADT), structure type, structural evaluation, waterway adequacy, and Strategic Highway Network (STRAHNET) designation plus several key geometric parameters.

- **Essentiality for public use**: Detour length, ADT, and STRAHNET designation.
- **Special reductions**: Detour length, traffic safety features, and structure type.

A complete description of the SR formula is available in *Recording and Coding Guide for the Structure Inventory and Appraisal of the Nation's Bridges*. [1]

The Health Index

To further its goal of preserving the bridge inventory, the California Department of Transportation has adopted a bridge health index as a performance measure. The health index is a single number indicator of the structural health of the bridge. This indicator is expressed as a value from 0 to 100 percent, corresponding to the worst and best possible conditions, respectively. The health index is calculated as a function of the fractional distribution of the bridge's element-level information across the range of applicable condition states, as shown in figure 4.

$HI = (\sum QCS_I \times WF_i)/(\sum TEQ \times We) \times 100\%$ {Worst = 0%, Best = 100%}

Figure 4. Equation. California Bridge Health Index [4].

Where:

HI = Health index.

QCS_i = Quantity in condition state i.

WF_i = Weighting factor for condition i.

TEQ = Total element value i.

We = Element indicator cost of other important indicator for each element.

Other State transportation departments also use health indices. The California Department of Transportation's index is cited here as an example.

Bridge Performance Measures Used in Other Countries

In the interest of understanding what other nations use to evaluate their bridge inventories, FHWA sponsored an international scan that resulted in the report *Bridge Preservation in Europe and South Africa.* [5] The following discussion of performance measures used in Finland, South Africa, and Sweden includes excerpts from that report.

Finland

Finland uses two performance indicators: one for repairs that improve a bridge's physical condition and one for rehabilitation, which encompasses improvements related to functional deficiencies. Details of these indicators are as follows:

- The repair index is a weighted combination of condition ratings that establishes priorities for repair projects and depends primarily on defect severity and ADT.
- The rehabilitation index responds to functional deficiencies and can indicate a need for improvement rather than repair.

The following discussion of performance measures used in Finland is representative of the findings in the international report and is excerpted directly:

> "Finland has a reference group of 106 bridges and 26 steel culverts. The performance of the group is closely monitored to improve knowledge of bridge behavior and durability, calibrate [bridge management system] deterioration models, and evaluate methods for field testing. Reference bridges are used in training and annual recertification of bridge inspectors…
>
> Finnra computes performance measures for defects, repair needs, and rehabilitation needs. A repair index is computed for the set of defects at a bridge. A rehabilitation index is computed for functional deficiencies. A repair index contributes to priorities for repair, unless the rehabilitation index indicates that a repair project should be set aside in favor of a rehabilitation project.
>
> Defects in a bridge are assigned ratings in each of four categories: weight (importance in the load path…), condition of the structural part (apart from this defect), urgency of the repair (rate of growth of defect), and damage

class (severity of the defect). For each bridge, a repair index, KTI, is computed for the set of defects, with the greatest weight placed on the worst defect. [See figure 5. Higher values indicate more important and urgent repairs.]

$$KTI = Max(Wt_i \times C_i \times U_i \times D_i) + k\sum(Wt_j \times C_j \times U_j \times D_j)$$

Figure 5. Equation. Finnra's Repair Index [5].

[Where:]
KTI = Repair Index.
Wt = Weight (importance) of the damaged structural part.
C = Condition of the structural part.
U = Urgency of the repair.
D = Class (severity) of damage.
k = A weighting factor for damage summation. The default value is 0.2.
i = Worst defect.
j = Other defects." [5]

The importance of the various structural parts is weighted from 0.20 for expansion joints to 1.00 for the superstructure. Condition points are assigned on a sliding scale as shown in table 4.

Table 4. Finnra Condition Rating. [5]

Condition Rating	Condition Points
0. New or like new	1
1. Good	2
2. Satisfactory	4
3. Poor	7
4. Very poor	11

Urgency of the necessary repair is classified with a point system of 10 points for repairs needed during the next 2 years, 5 points for repairs needed during the next 4 years, and 1 point for repairs not needed within 4 years. The severity of the deterioration or damage is also classified using a point system with 1 point for mild severity, 2 points for moderate severity, 4 points for serious severity, and 7 points for very serious severity. Finally, a weighting factor based on ADT is assigned for the essentiality of the repair, as shown in table 5.

Table 5. Finnra ADT Weighting Factor. [5]

ADT (vehicles /day)	Factor
> 6,000	1.15
3,000–6,000	1.10
1,500–3,000	1.00
350–1,500	0.90
< 350	0.85

Finnra's rehabilitation index, UTI, is calculated as shown in figure 6.

$$\text{UTI}=\text{kp}\times\text{kl}\times(\text{Condition}+\text{Load Capacity}+\text{Functionality})$$

Figure 6. Equation. Finnra's Rehabilitation Index [5].

Where:

kp = Factor for bridge total area.

kl = Factor for ADT.

The authors of the FHWA report also note, "The rehabilitation and reconstruction index, UTI, combines deterioration, bridge load capacity, and functionality to determine whether a bridge should be rehabilitated or replaced rather than repaired." [5] Finland sets annual goals for reducing severe deterioration in structures.

South Africa

In South Africa, performance measures are expressed as a condition index, I_c, for each defect and in terms of BCI for each bridge. The BCI combines the individual defect indices to provide an overall measure for the bridge. Figure 7 shows the formula used to calculate I_c.

$$I_c = 100\left[1 - \frac{(D+E)R}{32}\right]$$

Figure 7. Equation. South Africa's Condition Index [5].

Where:

D = Degree of defect.

E = Extent of defect.

R = Relevancy of defect.

D, *E*, and *R* are each assigned on a scale of 0–4, with 0 indicating no defect or that the defect is of no significance to the structural condition and 4 indicating the worst case in terms of severity, extent, or impact. The authors of the FHWA report note: "*Ic* equals 100 when there is no defect, and equals 0 when *D* and *E* and *R* are all equal to 4. A defect is critical if the *Ic* is below 40." [5] Figure 8 shows the formula for BCI.

$$BCI_n = \frac{(\sum_j Ic_j)ADT_n}{\sum_i ADT_i}$$

Figure 8. Equation. South Africa's BCI. [5] (Note: Due to formatting limitations, $\sum$ should be above *j* and above *i* in this equation, not beside them).

Where:

BCI_n = BCI for structure *n*.

Ic = Sum of condition index values for all relevant defects in structure *n*.

ADT_n = ADT for structure *n*.

ADT_i = Sum of values of ADT for all structures in the prioritization process.

BCI gives a combined indicator of the importance of defects and the importance of the bridge. A high BCI value indicates a good bridge, and a low value indicates a poor bridge. The further scaling by traffic volume will tend to increase BCI for heavily traveled bridges. The BCI for each bridge is computed, and a linear model is used for the decrease of the BCI with time.

The following discussion of performance measures used in South Africa is representative of the findings in the international report and is excerpted directly:

> "The index, *Ic*, is the dependent variable in the … deterioration model. Straight- line deterioration is proposed with the *Ic* declining at about 5 points per year. …
>
> Priorities for repairs are developed from two considerations. First is structural adequacy, as indicated by BCI values. Second is functional

importance, an evaluation computed from road class, bridge load capacity, detour length, etc. Generally, network-level optimization seeks the set of projects that offers the greatest reduction in defect relevancy, *R*, for a given budget.

Automated optimization yields a first list of repair projects. Next, projects for bridges are coordinated with projects for pavements in the same road section. Usually, a repair project at a bridge will attempt to remedy all relevant defects, not merely those with the highest priority values."[(5)]

Priorities for repairs in South Africa are determined by BCI and ADT.

Sweden

In Sweden, the preferred performance measure, lack of capital value (LCV), is defined as a fraction of the bridge replacement cost. The following discussion of performance measures used in Sweden is representative of the findings in the international report and is excerpted directly:

"Deterioration models consider structures in groupings determined by age and structural type. Deterioration is forecast as a continuing loss in capital value. The general form is [given by figure 9].

$$LCV = a_o + a_1 e^{rt}$$

Figure 9. Equation. Sweden's LCV [5].

[Where:]

LCV = Lack of capital value.

t = time.

a_o, a_1 , r = parameters of the model.

Parameters a_o, a_1, and r are specified by the user, or calibrated to historical trends in loss of capital value. This exponential model can represent both accelerating and decelerating change in capital value." [5]

The various formulae are used in different countries around the world for calculating bridge performance measures for individual bridges and for inventories of bridges. The measures can be calculated to describe the performance of an individual bridge relative to a fixed numeric scale, often 0–100. In each formula, the purpose is to calculate the sum of several weighted factors that relate to the performance of the bridge. Factors that are commonly used include the following:

- Condition of critical bridge components.
- Criticality of the components in the bridge structural system.
- Essentiality of the bridge to the highway network it serves.
- Cost of the element.

Essentiality of the bridge is often represented by volume of traffic carried or type of traffic carried (e.g., emergency responders, school vehicles, commercial traffic, military traffic, etc.).

When the performance measure is calculated for a group of bridges, the distribution of the performance measure values describes the performance of the group as a whole. These measures can be used to establish the order of priority for corrective or improvement actions or programs. Also, the bridge owner or agency can look at trends in bridge performance measures (rising or falling over time) and evaluate the effectiveness of bridge programs or the need to commit more resources to those programs. These formulae are very useful for their intended purposes.

SECTION 3. BRIDGE PERFORMANCE INDICATORS TO ADDRESS SPECIFIC CONCERNS

Highway bridges are normally expected to provide acceptable service for extended periods of time. Thus, most existing highway bridges have to be maintained in service for extended periods of time in spite of deterioration due to aggressive environmental stressors, aging of materials, and traffic increases. The safety and capacity of these bridges are highly influenced by their deterioration. In order to avoid the consequences of loss of serviceability or failure, maintenance programs are carried out by the responsible authorities. To make these programs cost-efficient, bridge life-cycle performance must be accurately predicted. [6-8] However, difficulties arise in the prediction of life-cycle performance because of the complexity and high uncertainty of deterioration mechanisms such as cracking, corrosion, and fatigue. Consequently, proper performance indicators are essential to evaluating bridge performance in a quantitative manner.

Many studies have focused on quantifying bridge performance by using deterministic, semi- probabilistic, and probabilistic indicators such as safety factors, partial safety factors, and the reliability index. Modern bridge design codes take uncertainty into account by including specific factors (i.e., load and resistance factors) in the computation of structural resistance and load effects with Load and Resistance Factor Design (LRFD) methods. However, prediction of bridge performance with deterioration requires the simultaneous use of several performance indicators. For example, the reliability index may be an adequate measure for quantifying the safety of a bridge component or system for ultimate capacity, but the redundancy index is required to evaluate the availability of warning before collapse.

Moreover, for bridges with deterioration and local damage, it is essential to consider performance indicators related to damage tolerance, such as reserve strength factor, residual strength factor, and vulnerability, together with the indicators for ultimate capacity. To obtain a desired bridge safety level, the values of performance indicators under consideration should not drop below prescribed minimum threshold levels.

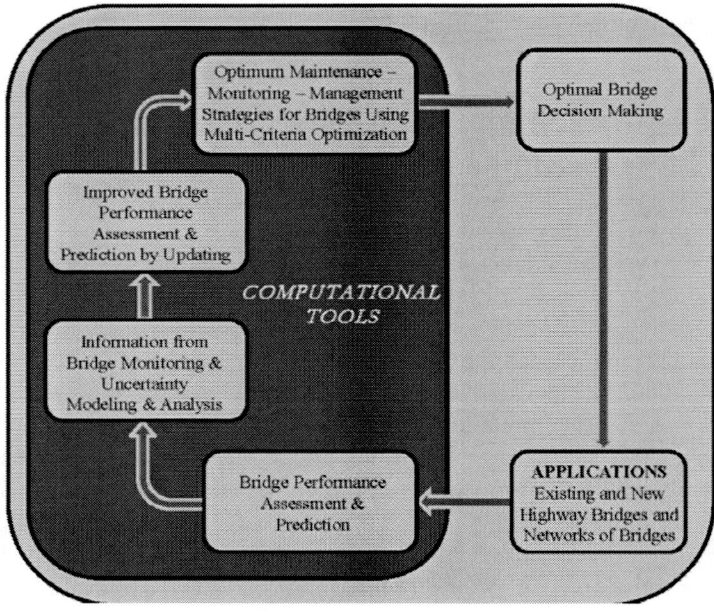

Figure 10. Chart. Integrated Life-Cycle Bridge Management Framework [12].

The life-cycle cost of bridges is another measure that decisionmakers have to balance with appropriate safety indicators. In recent years, life-cycle performance, safety, reliability, and risk of civil infrastructure systems have become emergent and key issues due to recurring natural and manmade disasters, the infrastructure crisis, sustainability issues, and global warming. Uncertainties are unavoidable in dealing with these problems. (See references 6 and 9–11.) Management of aging civil infrastructure involves significant expenditures. At a time of constrained public resources, difficult decisions are required to establish priorities for maintenance, rehabilitation, and replacement. Decisions regarding requirements for design, continued service, rehabilitation, or replacement should be based on multicriteria optimization under uncertainty, in order to balance conflicting requirements such as cost and performance. This can only be achieved through proper integrated risk management planning in a comprehensive life-cycle framework. Such a framework is shown in figure 10.

The purpose of this study is to review several performance indicators used to evaluate bridge performance. These indicators are briefly presented. In addition, a classification of bridge performance indicators is made depending on the approach (deterministic, semi-probabilistic (LRFD), or probabilistic) and level of concern (cross-section, component, or system level).

Structural Performance Indicators

Condition Rating (CR)

Condition ratings, *CR*, are used in the NBI to describe the existing, in-place bridge as compared to the as-built condition. Evaluation is for the materials, physical condition of the deck, superstructure, and substructure components of a bridge. Condition rating codes are properly used when they provide an overall characterization of the general condition of the entire component being rated. The load-carrying capacity is used in evaluating condition items. The fact that a bridge was designed for less than current legal loads and may be posted has no influence on condition ratings. The general condition ratings shown in section 2 are used as a guide in evaluating deck, superstructure, and substructure. The range of values possible for *CR* is shown in figure 11.

$$1 \leq \text{CR} \leq 9$$

Figure 11. Equation. Range of NBI Condition Ratings [1].

Condition State (CS)

Based on visual inspection, AASHTOWare Bridge Management™, the bridge management system of the American Association of State Highway and Transportation Officials (AASHTO), assigns condition states, *CS*, for bridge components. The condition states vary between 1 and 5, with each increasing condition state indicating a higher damage level. The range of values possible for *CS* is shown in figure 12.

$$1 \leq \text{CS} \leq 5$$

Figure 12. Equation. Range of AASHTOWare Bridge Management™ Condition States [13].

Margin of Safety (M)

The margin of safety, *M*, represents how much of the bridge cross-section, component, or overall bridge system capacity is held in reserve at a point in time. It can be expressed as shown in figure 13.

$$\text{M} = \text{R} - \text{Q}$$

Figure 13. Equation. Margin of Safety. [14]

Where:

R = Random variable representing the resistance effect.

Q = Random variable representing the load effect.

Time-Dependent Margin of Safety (M(t))

The time-dependent margin of safety, $M(t)$, is shown in figure 14.

$$M(t) = R(t) - Q(t)$$

Figure 14. Equation. Time-Dependent Margin of Safety [14].

Where:

t = Time.

R = Time-dependent variable representing the resistance effect.

Q = Time-dependent variable representing the load effects.

Probability of Failure (P_f)

Making the assumption that R and Q are statistically independent random variables, the instantaneous probability of failure, P_f, is shown in figure 15.

$$P_f = P(M(t) \leq 0) = \int_0^\infty F_R(x) f_Q(x) dx$$

Figure 15. Equation. Instantaneous Probability of Failure [15, 16].

Where:

R = Random resistance in a certain failure mode.
Q = Random load effect in the same failure mode.
$F_R(x)$ = Cumulative distribution function of R.
$f_Q(x)$ = Probability density function of load effect Q.

In many cases, it is impossible to evaluate P_f by analytical methods. Therefore, numerical methods such as the first-order reliability method, second-order reliability method, and Monte Carlo simulation are used.

Probability of Survival (Reliability) (P_s)

Failure and survival are complementary events. Therefore, the probability of survival, P_s, (also called reliability) is defined as shown in figure 16.

$$P_s = 1 - P_f$$

Figure 16. Equation. Probability of Survival [15, 16].

Reliability Index (β)

The reliability of a bridge can be expressed in terms of either P_f or its corresponding reliability index, β. For normal (Gaussian) distributed independent variables, β can be calculated as shown in figure 17.

$$\beta = \frac{E(R) - E(Q)}{\sqrt{\sigma^2(R) + \sigma^2(Q)}}$$

Figure 17. Equation. Reliability Index [15, 16].

Where:

$E(R)$ = Mean value of the resistance effect.

$E(Q)$ = Mean value of the load effect.
$\sigma(R)$ = Standard deviation of the resistance effect.
$\sigma(Q)$ = Standard deviation of the load effect.

For normally distributed independent variables, P_f and β are related as shown in figure 18.

$$P_f = 1 - \Phi(\beta)$$

Figure 18. Equation. Probability of Failure for Normally Distributed Independent Variables [15, 16].

Where:
$\Phi(\,)$ = Standard normal distribution function.
For the calibration of the Strength I limit state in *AASHTO LRFD Bridge Design Specifications*, $\beta = 3.5$ was used. [14]

Life-Cycle Cost

One of the most important measures in the evaluation of bridge performance is life-cycle cost. The proper allocation of resources can be achieved by minimizing the total expected cost while keeping structural safety at a desired level. The expected life-cycle cost can be expressed as shown in figure 19.

$$C_{ET} = C_T + C_{PM} + C_{INS} + C_{REP} + C_F$$

Figure 19. Equation. Expected Life-Cycle Cost [6].

Where:
C_{ET} = Expected life-cycle cost.
C_T = Initial design/construction cost.
C_{PM} = Expected cost of routine maintenance.
C_{INS} = Expected cost of performing inspections.
C_{REP} = Expected cost of repairs.
C_F = Expected cost of failure.

C_F is the cost of removal and replacement of an individual member or the cost of demolition of the existing bridge and replacement, if necessary, with a new bridge. Within this framework, all future costs are converted to their net present value. In the case of structural health monitoring (SHM), the expected life-cycle cost is as shown in figure 20, with figure 21 defining the expected cost of monitoring, C_{MON}.

$$C_{ET}^{0} = C_{T}^{0} + C_{PM}^{0} + C_{INS}^{0} + C_{REP}^{0} + C_{F}^{0} + C_{MON}^{0}$$

Figure 20. Equation. Expected Life-Cycle Cost with SHM [17, 18].

$$C_{MON} = M_{T} + M_{OP} + M_{INS} + M_{REP}$$

Figure 21. Equation. Expected Cost of Monitoring [17, 18].

Where:

M_T = Expected initial design/construction cost of the monitoring system.
M_{OP} = Expected operational cost of the monitoring system.
M_{INS} = Expected inspection cost of the monitoring system.
M_{REP} = Expected repair cost of the monitoring system.

Using this approach, the benefit of the monitoring system, B_{MON}, can be determined through a comparison of the expected life-cycle total cost with and without monitoring, as shown in figure 22.

$$B_{MON} = C_{ET} - C_{ET}^{0}$$

Figure 22. Equation. Benefit of the Monitoring System [17, 18].

Using figure 22, monitoring is economically beneficial if $B_{MON} > 0$. A monitoring benefit may be realized through better treatment of maintenance and repair activities as well as a lower level of risk over the life of the structure. Bridge managers can prevent or reduce adverse consequences by using monitoring data.

Safety Factor in Allowable Stress Design (SF)

Allowable Stress Design (ASD) is based on the concept that the maximum stress in a component should not exceed a certain allowable stress under normal service conditions. The limiting stress, which can be yield stress or stress at instability or fracture, is divided by a safety factor to provide the allowable stress. The safety factor, *SF*, is used to provide a design margin over the theoretical design capacity and is defined as shown in figure 23.

$$SF = \frac{\sigma_u}{\sigma_{all}}$$

Figure 23. Equation. Safety Factor in ASD [19].

Where:

σ_u = Maximum usable stress, which can be the yield stress, buckling stress, or ultimate stress.

σ_{all} = Allowable stress.

Partial Factors Used in LRFD (ϕ , γ)

In LRFD, resistance *R* and load *Q* are considered statistically independent random variables. If *R* is greater than *Q*, a margin of safety exists. On the other hand, since *R* and *Q* are random variables, there is a probability that *R* is smaller than *Q*. The typical inequality for safety checks in LRFD is shown in figure 24.

$$\phi R_n \geq \gamma_D Q_{Dn} + \gamma_L Q_{Ln} + \ldots$$

Figure 24. Equation. Inequality for Safety Check in LRFD [14].

Where:

ϕ = Resistance factor associated with nominal resistance R_n.

γ_D, γ_L = Partial load factors associated with the dead and live loads effects Q_{Dn} and Q_{Ln}.

In LRFD, resistance R_n is reduced by resistance factor ϕ, and loads are amplified or reduced by load factors.

Load Modifier Factor Used in LRFD (η_i)

In *AASHTO LRFD Bridge Design Specifications*, an additional load modifier factor, η_i, relating to ductility, redundancy, and operational importance is used, as shown in figure 25. [14]

$$\phi R_n \geq \sum \eta_i \gamma_i Q_i$$

Figure 25. Equation. Load Modifier Factor in LRFD for Ductility, Redundancy, and Operational Importance [14].

In figure 25, η_i, is defined by figure 26 for loads for which a maximum value of γ_i is appropriate and by figure 27 for loads for which a minimum value of γ_i is appropriate.

$$\eta_i = \eta_D \eta_R \eta_I \geq 0.95$$

Figure 26. Equation. Load Modifier Factor for Maximum Value of γ_i. [14].

$$\eta_i = \frac{1}{\eta_D \eta_R \eta_I} < 1.0$$

Figure 27. Equation. Load Modifier Factor for Minimum Value of γ_i. [14].

Where:

η_D = Factor relating to ductility.
η_R = Factor relating to redundancy.
η_I = Factor relating to operational importance.

Collapse Load Multiplier (λ)

In the plastic analysis of structures, collapse load multiplier, λ, is a theoretical factor by which a set of loads acting on the structure must be multiplied just enough to cause the structure to collapse. The load can be taken as the service load conditions, and the strength of the structure can be determined from idealized plastic material strength properties. Three main loading histories are considered.

Proportional loading implies that the applied load can be defined at all stages by a single parameter, λ, which amplifies the service load, $Q = Q_D + Q_L$, so that the ultimate load, Q_U, is as shown in figure 28.

$$Q_U = \lambda_0 (Q_D + Q_L)$$

Figure 28. Equation. Collapse Load Multiplier for Proportional Loading [20].

This extension of ASD practice is used in the plastic design of steel structures but is an unrealistic concept because dead loads are not subject to the same variations as live loads.

Combined loading assumes that the dead load is fixed and the live load only is variable, as shown in figure 29.

$$Q_U = Q_D + \lambda_L Q_L$$

Figure 29. Equation. Collapse Load Multiplier for Combined Loading [20].

Arbitrary loading assumes that the dead and live loads vary independently in both magnitude and time. The dead load first reaches its full factored value, $\lambda_D Q_D$, before the live load is applied from zero to its full value, $\lambda_L Q_L$, as shown in figure 30.

$$Q_U = \lambda_D Q_D + \lambda_L Q_L$$

Figure 30. Equation. Collapse Load Multiplier for Arbitrary Loading [20].

The inequality $R_U \geq Q_U$ has to be satisfied at the limit, where R_U is the plastic resistance of the structure.

Return Period (T)

The loads due to natural phenomena such as earthquakes, storms, and high winds have randomness not only in space but also in time. The randomness in time can be considered in terms of return period, T. Return period is an average duration between consecutive occurrences of an event. It should be noted that the actual time between the occurrences is T, which is a random variable, and T is only average duration. T can be expressed as shown in figure 31.

$$\bar{T} = E(T) = p(1 + 2q + 3q^2 + \ldots)$$

Figure 31. Equation. Return Period for Loads Due to Natural Phenomena [21].

Where:

p = Probability of occurrence of the event.

q = Corresponding probability of nonoccurrence (therefore, $q = 1 - p$).

The infinite series inside the parentheses yields $1/p^2$. Therefore, the average duration between consecutive occurrences of an event is as shown in figure 32.

$$\bar{T} = \frac{1}{p}$$

Figure 32. Equation. Average Duration Between Consecutive Occurrences of an Event [21].

Existing bridge design codes use different return periods for different hazards. For example, the calibration of the live load factors in *AASHTO LRFD Bridge Design Specifications* is based on the 75-year maximum load effect, while a 1,000-year return period has been proposed for seismic hazards, and 100- and 500-year flood levels are used for scour. [14]

Cumulative Time Probability of Failure (F(t))

The probability of failure within a certain period of time is called the cumulative time probability of failure, *F(t)*. There are two approaches for computing cumulative probability of failure.

In the first approach, *F(t)* is computed considering only one random variable, the time T_f at which the component or system fails. Probability concepts are applied to compute *F(t)*. In this approach, *F(t)* up to time t_f can be calculated as shown in figure 33.

$$F(t_f) = P(T_f \leq t_f) = \int_0^{t_f} f(u)du$$

Figure 33. Equation. Cumulative Time Probability of Failure, One Random Variable [22].

Where:

$F(t_f)$ = Area under the probability density function $f(u)$ of the time to failure from t_0 to t_f.

In the second approach, *F(t)* is computed considering changes in both time-variant resistance and load. The cumulative time probability of failure of a component whose resistance is deteriorating subjected to time-variant load is shown in figure 34.

$$F(t) = 1 - \int_0^{\infty} \exp\left[-\lambda t\left[1 - \frac{1}{t}\int_0^{t_L} F_S\{r \cdot g(\xi)\}\right]\right] f_{R_0}(r)dr$$

Figure 34. Equation. Cumulative Time Probability of Failure, Time-Variant Resistance and Load [22].

Where:

λ = Mean occurrence rate of stochastic load event in a Poisson process.

$F_s(\ )$ = Cumulative time distribution function used to define the load intensity.

$f_{Ro}(r)$ = Probability density function of the initial resistance R_0.

$g(t)$ = Mean value of $G(t)$.

$G(t)$ = Resistance degradation function.

Enright and Frangopol extended the formulation in figure 34 and determined *F(t)* of weakest-link and fail-safe systems with applications to bridges.(23,24)

Cumulative Time Probability of Survival (S(t))

Cumulative time probability of survival, *S(t)*, also called survival function, is the probability that a component or system survives until time *t*. It is equal to the reliability function, which is the probability that a component or system is still functioning at time *t*. *S(t)* is the complement of the *F(t)* and can be expressed as shown in figure 35.

$$S(t) = 1 - F(t) = P(T > t) = \int_{t_f}^{\infty} f(u)du$$

Figure 35. Equation. Cumulative Time Probability of Survival [22].

Where:

$S(t)$ = Area under the probability density function $f(u)$ of the time to failure of t_f to infinity.

Reserve Strength Factor (R_1)

Reserve strength factor, R_1, is defined as the ratio of the load-carrying capacity of the intact structure (or component), *C*, to the applied load on the structure (or component), *Q*, as shown in figure 36.

$$R_1 = \frac{C}{Q}$$

Figure 36. Equation. Reserve Strength Factor. (See references 25–29).

R_1 varies from a value of infinity, when the intact structure (or component) has no load, to a value of 1.0, when the nominal load on the intact structure (or component) equals its capacity.

Residual Strength Factor (R_2)

The residual strength factor, R_2, provides a measure for the strength of the system in a damaged condition compared to the intact system. It is defined as the ratio of the capacity of the damaged structure (or component), C_d, to the capacity of the intact structure (or component), C, and can be expressed as shown in figure 37.

$$R_2 = \frac{C_d}{C}$$

Figure 37. Equation. Residual Strength Factor. (See references 25–29).

R_2 takes values between 0, when the damaged structure has zero capacity, and 1.0, when the damaged structure does not have any reduction in load-carrying capacity.

Redundancy Factor (R_0)

Redundancy factor, R_0, is defined as shown in figure 38.

$$R_0 = \frac{1}{1 - R_2}$$

Figure 38. Equation. Redundancy Factor. (See references 25–29).

R_0 varies between 1.0, when the damaged structure has zero capacity, and 0, when the damaged structure does not have any reduction in load-carrying capacity.

Damage Factor (D)

Damage factor, *D*, is used to represent the loss in cross-sectional area of a bridge component. It is defined as shown in figure 39.

$$D = 1 - \frac{a_d}{a}$$

Figure 39. Equation. Damage Factor [19, 20].

Where:

a_d= Cross-sectional area of the damaged portion of the bridge component.
a = Intact cross-sectional area of the bridge component.

Redundancy Index (RI)

Redundancy, which is a measure of reserve capacity, can be defined as the availability of warning before structural collapse occurs. The failure of a single member will not cause failure of a redundant structure. There are several measures for probabilistic redundancy index, *RI*, including those shown in figure 40 through figure 42.

$$RI_1 = \beta_{\text{intact}} - \beta_{\text{damaged}}$$

Figure 40. Equation. Probabilistic Redundancy Index 1 [25, 30, 31] .

$$RI_2 = \frac{\beta_{\text{intact}}}{\beta_{\text{intact}} - \beta_{\text{damaged}}}$$

Figure 41. Equation. Probabilistic Redundancy Index 2 [25, 30, 31].

$$RI_3 = \frac{P_{f(dmg)} - P_{f(sys)}}{P_{f(sys)}}$$

Figure 42. Equation. Probabilistic Redundancy Index 3 [25, 30, 31].

Where:

β_{intact}= = *RI* of the intact system.
$\beta_{damaged}$= = *RI* of the damaged system.
$P_{f(dmg)}$= Probability of damage occurrence to the system.
$P_{f(sys)}$ = Probability of system failure.

To account for member ductility, system redundancy, and operation importance, *AASHTO LRFD Bridge Design Specifications* applies a load modifier factor η_i relating to ductility (η_D), redundancy (η_R), and operational importance (η_I) (see figure 27). [14]

Time-Variant Redundancy Index (RI(t))

Frangopol and Okasha investigated several time-variant redundancy indices, *RI(t)*, based on point-in-time β and P_f. It was shown that the redundancy indices shown in figure 43 and figure 44 are most consistent. [32, 33]

$$RI_1(t) = \frac{P_{y(sys)}(t) - P_{f(sys)}(t)}{P_{f(sys)}(t)}$$

Figure 43. Equation. Time-Variant Redundancy Index 1 [32, 33].

$$RI_2(t) = \beta_{f(sys)}(t) - \beta_{y(sys)}(t)$$

Figure 44. Equation. Time-Variant Redundancy Index 2 [32, 33].

Where:

$P_{y(sys)}(t)$ = System probability of first yield at time *t*.

$P_{f(sys)}(t)$ = Probability of system failure at time *t*.

$\beta_{y(sys)}(t), \beta_{f(sys)}(t)$ = Reliability indices with respect to first yield and system failure at time *t*.

An increase in the value of *RI* indicates a higher system redundancy. A structural system is considered non-redundant if *RI* = 0.

Vulnerability (V)

Vulnerability, *V*, is one of the key measures used to capture the essential feature of damage-tolerant structures. A probabilistic measure of *V* can be defined as the ratio of the failure probability of the damaged system to the failure probability of the undamaged system, as shown in figure 45.

$$V = \frac{P(r_d, Q)}{P(r_0, Q)}$$

Figure 45. Equation. Vulnerability [34].

Where:

r_d = A particular damaged state.

r_0 = An undamaged system state.

Q = Prospective loading.

$P(r_d, Q)$ = Probability of failure of the system in the damaged state.

$P(r_0, Q)$ = Probability of failure of the system in the pristine state.

V refers to vulnerability of the system in state r_d for prospective loading Q. The value of V is 1.0 if the probabilities of failure of the damaged and intact systems are the same.

Damage Tolerance (D_t)

Damage tolerance, D_t, can be defined as the reciprocal of V, as shown in figure 46.

$$D_t = \frac{1}{V} = \frac{P(r_0, Q)}{P(r_d, Q)}$$

Figure 46. Equation. Damage Tolerance [34].

Ductility (Δ)

Ductility, Δ, is generally defined as the ability of a bridge component or the entire bridge to sustain large deformations without collapse. A ductility index could be defined as shown in figure 47.

$$\Delta = \Delta_c - \Delta_{el}$$

Figure 47. Equation. Ductility [34].

Where:

Δ_c = Deformation at collapse.

Δ_{el} = Deformation associated with the limit of elastic range.

Robustness (RO)

Robustness, *RO*, is one of the key measures in the field of progressive collapse and damage-tolerant structures. Although robustness is recognized as a desirable property in structures and systems, there is not a widely accepted theory on robust structures. [35] In general, *RO* is defined as insensitivity of the safety of a structure to local failure or the ability of a structure to prevent failure progression.

Table 6. Classification of Structural Performance Indicators

Performance Indicator	Notation	Approach			Level		
		Deterministic	Semi-Probabilistic (LRFD)	Probabilistic	Section Level	Component Level	System Level
Condition rating	CR	✓				✓	
Condition state	CS	✓				✓	
Margin of safety	M			✓	✓	✓	✓
Time-dependent margin of safety	$M(t)$			✓	✓	✓	✓
Probability of failure	P_f			✓	✓	✓	✓
Probability of survival	P_s			✓	✓	✓	✓
Reliability index	β			✓	✓	✓	✓
Life-cycle cost	LCC			✓			✓
Safety factor in ASD	SF	✓			✓	✓	
Partial load factors in LRFD	ϕ, γ		✓		✓	✓	
Load modifier factor in LRFD	η_i		✓			✓	✓
Collapse load multiplier	λ	✓					✓
Return period	T		✓	✓	✓	✓	✓
Cumulative time probability of failure	$F(t)$			✓		✓	✓
Cumulative time probability of survival	$S(t)$			✓		✓	✓
Hazard rate	$h(t)$			✓		✓	✓
Cumulative hazard rate	$H(t)$			✓		✓	✓
Reserve strength factor	R_1	✓				✓	✓
Residual strength factor	R_2	✓				✓	✓
Redundancy factor	R_0	✓					✓
Damage factor	D	✓			✓		
Redundancy index	RI			✓			✓
Time-variant redundancy index	$RI(t)$			✓			✓
Vulnerability	V			✓			✓

Table 6. (Continued)

Performance Indicator	Notation	Approach			Level		
		Deterministic	Semi-Probabilistic (LRFD)	Probabilistic	Section Level	Component Level	System Level
Damage tolerance	D_t			✓			✓
Ductility	Δ	✓				✓	✓
Robustness	RO			✓			✓
Resilience	RE			✓			✓
Risk	$\mathcal{R}$			✓			✓

✓ Checkmark indicates performance indicator meets classification; blank cells indicate performance indicator does not meet classification.

Resilience (RE)

Resilient structures respond well to extreme events. They reduce the probabilities of failure, the consequences of failure, and the time for recovery. Resilience, *RE*, can be measured by the functionality of an infrastructure system after a disaster and by the time it takes for a system to return to predisaster levels of performance. [36] Despite several *RE* measures proposed in the literature, a standard measure for bridge resilience has yet to be specified.

Risk ($\mathcal{R}$)

Risk, $\mathcal{R}$, may be expressed as a function of the probability of occurrence of adverse event *A*, *P*(*A*), and the consequence of this event, *K* (typically expressed in monetary terms). Often, $\mathcal{R}$ is defined as the product of *P*(*A*) and *K*, as shown in figure 48.

$$\Re = P(A) \bullet K$$

Figure 48. Equation. Risk [34].

The uncertainties in both *P*(*A*) and *K* will carry over in calculating $\mathcal{R}$.

Minimizing risk is one of the main objectives of bridge management. This can be achieved by minimizing the probability of occurrence of the adverse event (e.g., probability of bridge collapse), minimizing the consequences associated with this event, or minimizing both. Low-probability, high-consequence events are of particular relevance in risk-informed decisionmaking and management of aging bridges.

Classification of Structural Performance Indicators

A classification of the defined structural performance indicators depending on the approach (deterministic, semi-probabilistic (LRFD), or probabilistic) and level of concern (cross section, component, or system level) is presented in table 6. More than one checkmark in the same category indicates that the performance measure can be classified in each section depending on the situation. Risk-informed assessment and management of the highway bridge infrastructure in the United States requires a set of reliability-based performance indicators and decision tools. Consideration of multiple performance indicators in the evaluation of structural performance is inevitable. The indicators that should be considered depend on the priorities and objectives of the decisionmakers. The main remaining challenge lies in the implementation of reliability-based performance indicators that account for the presence of both natural randomness (i.e., aleatory uncertainty representing the natural variability or randomness of nature) and imperfect knowledge (i.e., epistemic uncertainty representing our imperfect ability to model reality) in bridge engineering. The indicators for evaluating bridge performance are not limited to those mentioned in this study.

SECTION 4. IMPROVING BRIDGE PERFORMANCE

Keys to Improving Bridge Performance

Bridges are a critical part of the transportation system. The bridge infrastructure is a vast and valuable asset that must be properly managed in the interests of efficiency, safety, economy, national security, and protection of local and global environments. The need for useful, reliable bridge performance measures is clear. Additionally, there is a need for methods to evaluate the impact of different scenarios of funding, maintenance practices and priorities, design methodologies, and new technologies on future bridge performance. The performance measures described in section 2 are consistently applied and well-tested in practice. However, they do not lend themselves to predictive efforts nor to analysis of various "What if … ?" scenarios. The LTBP Program will concentrate on collecting information and data that will allow exploration of scenarios that have actually happened. This will lead to tools such as better deterioration models, allowing better predictions of bridge performance.

The performance indices described in this report generally attempt to address the bridge as a whole entity and aggregate values among a population of bridges. However, bridges are composed of several unique elements that work as a system. Each of these elements responds to a different set of factors that govern its performance. Even when the same factor affects the performance of more than one element, the manner and degree to which that factor impacts each element may vary considerably. Many aspects of bridge performance are not well understood, and current indices of bridge performance are usually based on objectives that are imprecise and data that are not consistent or well documented. Many attempts at assessment of performance (of the complete structure or key elements) rely on expert opinion and significant assumptions and generalizations. Yet, decisions at many different levels and for many important purposes are based on these performance assessments.

The keys to improving bridge performance measures are as follows:

- Establishing clear, objective measures that are relevant to different levels of decisionmaking.
- Identifying the elements and characteristics of a bridge that most seriously impact the bridge in the four performance categories: condition and durability, functionality, structural integrity and safety, and cost. This exercise should consider both short-term issues and the long-term potential for improved performance through innovative materials, enhanced inspection technology, improved design concepts, and evolving construction and maintenance methods.
- Identifying critically needed data for experimental studies that will improve the knowledge of the multivariable cause-and-effect relationships that govern performance degradation.
- Collecting data to fill gaps and create valid models that describe deterioration mechanisms, predict future deterioration, and support more realistic life-cycle cost calculations. Such information can then be used to further calculate operational efficiency and safety versus various levels of condition and capacity, as well as the probability of structural system failure at both the service and ultimate-limit states.
- Improving the relevance and quality of the data collected and used to support analyses and calculate performance indicators.
- Developing tools and systems for calculating and disseminating results about performance indicators.

Better Data on Bridge Performance in the United States

The primary sources of data on bridges are the NBI and bridge management databases used by many States. Many transportation departments also have databases for recording bridge maintenance actions and for management and planning purposes. The NBI was created to fill the knowledge gap in bridge inventory and condition information but not to support the assessment of bridge performance over time. Various parties have been able to use the NBI to produce results and products that have implications for the assessment of bridge performance. When using NBI data to assess bridge performance, the following characteristics must be considered:

- The NBI contains condition ratings for the major structural elements of the bridge—deck, superstructure, and substructure—plus channel, channel protection, and culverts. The NBI does not contain data on the condition of the myriad individual subelements of a bridge such as beams, pier columns, and abutments.
- The NBI contains appraisal ratings for a few key features of the bridge, most notably structural evaluation, deck geometry, and scour criticality.
- The NBI does not contain quantitative measurements of differing conditions or the locations of differing conditions with respect to the geometry of the element being inspected.
- In most cases, the condition ratings are assigned based solely on visual inspections. Underlying causes of damage, such as rebar corrosion, are not identified until surface damage appears. NBI ratings do not reflect incipient degradation.
- The full range of codes (from 9 for excellent condition to 0 for failed condition) consists of discrete integers corresponding to the guiding language. This type of data is not readily amenable to rigorous mathematical analysis with the purpose of charting continuous change or for predicting future changes.

The most common bridge management system in the United States is the AASHTOWare™ Bridge Management software. The developers of this system incorporated a detailed condition assessment approach. An element-level inspection system was developed to track both the severity and the extent of different problems. AASHTOWare™ Bridge Management addresses deterioration as a probabilistic, rather than deterministic, process and is able to automatically update previous deterioration predictions as more cycles of historical inspection data are input. The initial probabilities of transition from one condition state to the next were determined from consensus of expert opinion. These can be modified by the licensing agency if desired. Under the element-level inspection system, condition data are recorded on the individual elements of the bridge rather than on the general elements of deck, superstructure, and substructure. This expands the data collected while allowing the use of specific guidance and employing precise engineering language for inspectors to rate the elements. Thus, the severity of any deterioration is defined, and the extent is estimated and recorded. While the element-level data system provides more granular information than the NBI condition ratings, it is still based mostly on visual inspection of surface conditions and relies on inspector training and experience for data quality. It does not, for the most part, document latent defects, such as initial corrosion of reinforcing bars that could soon change the observed condition at the surface level. The NBI and element-level databases can be used as a fundamental resource in evaluating bridge performance. The past, current, and future data contained in these two resources are helpful for the following:

- Identifying trends in bridge performance.
- Identifying general parameters that govern performance.
- Identifying representative bridges and service conditions that can provide a real-life laboratory for studying performance issues.

Bridge owners may also possess other useful data, such as the following:

- Design drawings and specifications.
- Analytical models.
- Construction records.
- Inspection reports.
- Photographic documentation.
- History of maintenance actions and timing.
- Financial records of maintenance costs.

Beyond these current resources, the data needed to properly evaluate the important aspects of bridge performance can be quite extensive. Table 7 and table 8 show the breadth of data that may be necessary to better understand bridge performance in the four main categories.

The steps to obtaining better data for performance assessments are as follows:

1. Isolate specific aspects of bridge performance that have the potential to significantly increase the costs associated with bridges (these costs may be in terms of dollars for bridge maintenance and rehabilitation; damage to the environment; property damage, injuries, or fatalities in bridge-related accidents; or delays associated with congestion at narrow bridges or bridges under repair).
2. Establish working hypotheses as to the cause of poor performance.
3. Identify the parameters that have the most impact on performance.
4. Identify the critical data needed to prove or disprove the hypotheses.
5. Identify the data collection methodologies that best balance effectiveness and data quality with the costs of data collection. The focus should be on obtaining the best available data rather than on trying to obtain perfect data.

This process will support the ability to design, implement, and complete statistically sound experimental studies to prove or disprove the hypotheses. The result will be better understanding of bridge performance and the knowledge necessary to implement policies, programs, and specific actions that will result in improved performance. An example of how this process would work for the issue of performance of FO bridges is described in the following section.

Table 7. Durability and Serviceability Performance Data. Category

Category	Data
Design and construction	Design plans and specifications
	Critical design details
	Change orders
	Inspection notes
	Construction quality assurance and quality control
	Corrosion protection measures
Operating conditions	Local climate
	Snow and ice removal practices
	Freeze-thaw cycles
	Rainfall and runoff; drainage control
	Marine environment
	Industrial pollutants
Dynamic loadings	Traffic volume
	Truck volumes and weights
	Weigh-in-motion data
	Overload permits
	Debris, ice
	Impact loads
	Flexibility, vibrations
Corrosion protection measures	Concrete cover over reinforcement
	Corrosion resistant reinforcement
	Deck overlays, membranes, and sealers
	Other concrete sealers
	Steel coatings—high-performance or weathering steel
	Concrete qualities—high-performance concrete
Material conditions	Concrete
	Steel
	Reinforcing bars
	Prestressing steel
	Deck
	Concrete superstructure
	Steel superstructure
	Concrete substructure
Geometric data	Deflections
	Rotations
	Settlements
	Loss of camber
	Horizontal alignment and skew
Components	Bearings
	Joints
	Approach slabs
	Details requiring inspection by non-destructive evaluation

Table 8. Functionality, Cost, Structural Integrity, and Organizational Data

Categories		Data
Functionality: User safety and service	Operational efficiency	Traffic volumes
		Congestion and delay times
		Safety hazards
		Accident rates and types
	Network-level performance	Route redundancy
		Detour lengths and costs
		Other bridges in the corridor
	Intersection/interconnection	Military route
		Multimodal interconnections
		Accommodation of other infrastructure
	Societal/environmental impacts	Fuel usage
	Environmental issues	Air quality
		Water quality
		Toxic wastes
Costs: Agency and user	Original construction	Design costs
		Construction costs
	Life-cycle costs	Inspection and condition assessment
		Preservation
		Maintenance, repairs
		Rehabilitation
		Demolition, removal, and disposal
		Work zone maintenance of traffic
Structural integrity: Safety and stability	Safety and stability	Global
		Member
		Redistribution
		Resilience
		Structural redundancy
	Extreme events	Foundation type
		Accident risks—fire, impact
		Soil and hydraulic conditions
		Scour vulnerability
		Scour mitigation measures
		Seismicity
		Seismic design considerations
		Hazard return periods
Organizational issues		Asset management policies
		Revenue sources
		Distribution decision authority
		Organizational structure and culture
		Knowledge management
		Human resources
		Quality of education
		Incentives for growth

Example: Performance of FO Bridges

Objective

The objective of the study is to examine the performance of bridges classified as FO and evaluate the frequency and severity of negative impacts (e.g., accidents, vehicle-bridge collisions, and instances where the bridge route does not function adequately) as a result of alignment disparities.

What Effect Does Inadequate Roadway Alignment Have on Road Safety and Traffic Flow?

Hypotheses: The following hypotheses are developed concerning inadequate roadway alignment:

- A moderate to severe misalignment of the approach roadway and the bridge roadway may cause momentary reductions in speed and erratic movements by drivers.
- These actions may contribute to accidents close to or on the bridge. They may also contribute to impaired traffic flow.

Data: Data requirements for this study are as follows:

- Geometry of approach roadway and bridge roadway:
    - Number of traffic lanes.
    - Lane widths.
    - Shoulder widths.
    - Approach roadway alignment.
- Functional class of highway.
- Auto and truck traffic volumes.
- One-way or two-way traffic.
- Traffic control features at and near the bridge.
- Posted speed and typical traffic speeds.
- Current and past accident data at or on the bridge:
    - Type.
    - Severity.
    - Documented causes.
- Ambient conditions such as light and precipitation.

The effort to analyze and better understand various aspects of bridge performance should not be confused with the need to continue collecting required NBI data and the element-level data now collected to support bridge management systems. No recommendations are being made to expand the volume or enhance the quality of the data currently collected for these purposes.

Better Understanding of Cause and Effect

Bridge performance measures should be able to provide a link between a specific parameter (or set of parameters) and the level of some desirable quality or characteristic of bridges or bridge elements. For example, a traffic safety measure should provide a link between accident experience and geometric characteristics such as bridge width. However, the performance of any single bridge element is dependent on complex interactions of multiple and often interrelated factors, including the original design parameters and specifications (bridge type, materials, geometries, and load capacities); the initial quality of materials and of the as-built construction; varying environmental conditions of climate and air quality; corrosion and other deterioration processes; traffic volumes and percentage of truck traffic; and the type, timing, and effectiveness of preventive maintenance, minor and major rehabilitation actions, and ultimately, replacement actions. Realistically, it is usually impossible to isolate one specific parameter as the governing parameter for a specific quality. Even in the simple example of accident history, several different factors are linked to safety, including bridge and roadway geometry, traffic volumes and vehicle types, travel speeds, weather conditions and winter maintenance operations, night visibility, and pavement and deck riding surface condition.

Analysis of the 2011 NBI data shows that the primary cause for a bridge to be flagged as SD or FO is related to deck geometry—a bridge is FO because the roadway width is considered too narrow for the traffic volumes currently using the bridge. [2] Figure 49 provides the top 10 reasons why a bridge is rated as deficient.

Calculation of Federal SR for a given bridge, particularly the part of SR dependent on functional characteristics, is based on somewhat arbitrary definitions and significant assumptions. For bridges whose geometry does not meet currently accepted standards, many of the criteria used to calculate SR are not well documented and may not be well understood by bridge engineers. Traffic safety in the vicinity of bridges is an important parameter of bridge performance, yet there are no proven relationships that relate safety to bridge characteristics such as clear deck width, clearances, or approach roadway alignment. The *Recording and Coding Guide for the Structure Inventory and Appraisal of the Nation's Bridges* presents a table for evaluating deck geometry that considers traffic volumes, lane widths, direction of traffic, and type of highway system. [1] No apparent research supports these numbers, and the method ignores possibly causative or complicating factors such as climatic conditions, percent of trucks in the traffic stream, approach roadway alignment, and posted speed.

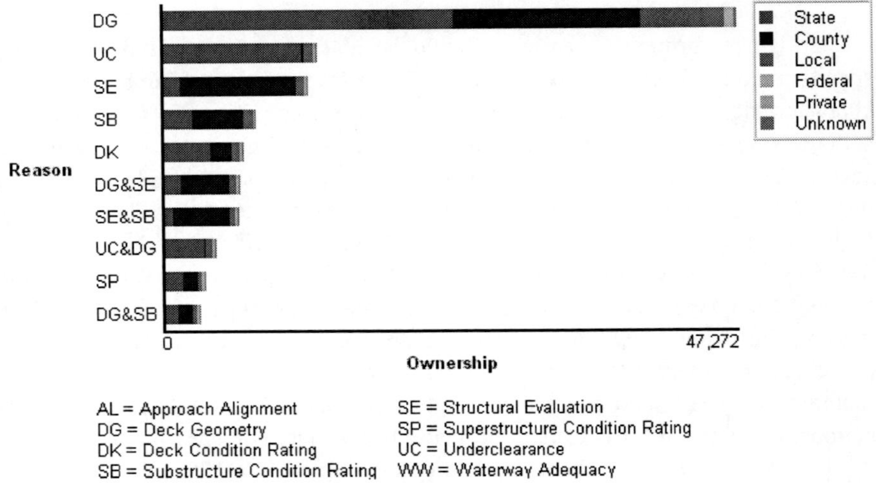

Figure 49. Graph. Reasons for Bridge Rating of SD or FO [2].

Therefore, a study to examine the cause-and-effect relationships between bridge and traffic parameters and accident history near and on bridges would be useful. The data in the NBI and other available databases can be mined to create a subset of bridges that reflects a range of the key variables, including traffic volumes, truck traffic, deck widths, and perhaps other causative factors such as climatic conditions, approach roadway alignment, and posted speed.

The FHWA Bridge Management Information Systems Laboratory was established to identify and analyze causes and trends of deficiencies within the Nation's bridge inventory. The laboratory has developed or acquired the tools to support sophisticated analytical research on existing disparate data sources through a geographical information system platform combined with relational database management systems software and advanced mathematical and statistical software. Under a program such as LTBP, data about type and frequency of accidents can be monitored, and analysis of long-term accident experience can be used to correlate accident potential with the most relevant bridge parameters. Ultimately, the findings could be used to better understand the bridge deficiencies that most affect functional characteristics of bridges.

The preceding discussion is one example of an aspect of bridge performance that could be improved by a better understanding of cause-and-effect relationships supported by accurate, reliable data. This principle of establishing and proving cause-and-effect relationships using high-quality research data is one of the major goals of the LTBP Program. With more relevant and reliable data and improved understanding of key cause-and-effect relationships, the knowledge necessary to improve bridge performance can be obtained.

Critical Aspects of Bridge Performance

As previously noted, it is possible to characterize bridge performance issues under four broad categories: structural condition, functionality, structural integrity, and costs to the user and the agency. Within each of these categories are specific issues that may be considered higher priorities than others, including consideration of one or more of the following:

- Cost to the owner agency in terms of inspection, engineering, maintenance rehabilitation, preservation, and replacement.
- Risk associated with failure of the bridge or a critical element.
- Property damage, injuries, and fatalities associated with accidents.
- Delay time and detour costs for commuters, commercial shippers, and tourists due to restricted conditions near or on a bridge.
- Excess fuel consumption and emissions due to inefficient or extended driving times.

Because owner agency resources are often severely limited, it is necessary to examine each issue of bridge performance and weigh the costs of poor performance against the costs of actions necessary to improve performance. This assumes that the causes of poor performance have been determined and solutions are available.

Under the LTBP Program, research has been done on evaluating the highest priorities in bridge performance. A consensus has emerged based on expert opinion from owner agencies and other bridge experts. The future results of the LTBP Program will provide knowledge to assist owners in properly making necessary analyses. High-priority performance issues will be outlined in a forthcoming program report, which will identify specific research to be undertaken in terms of objectives, key questions, working hypotheses, and critical data needed.

CONCLUSION

One simple definition of *performance* is that performance equates to accomplishment of a specified purpose or set of purposes. For the purposes of the LTBP Program, the following definition is used: Bridge performance encompasses how bridges function and behave under the complex and interrelated factors they are subjected to day in and day out—traffic volumes, loads, deicing chemicals, freeze-thaw cycles, rains, or high winds. Bridge design, construction, materials, age, and maintenance history also play roles in performance. Virtually everyone in the United States, from bridge maintenance engineers to the everyday road user, has a stake in ensuring that the performance of bridges is good or excellent in terms of durability, operational capacity, roadway safety, resistance against failure, environmental neutrality, and life-cycle costs. Each group needs to have confidence in their understanding of the indicators by which they measure bridge performance. For the everyday commuter or casual user, an indicator such as Federal SR may suffice. For commercial interests, shippers, and drivers, a simple measure of bridges with posted weight limits or geometrical dimensions may suffice. These groups also need assurance that satisfactory bridge performance is being provided at reasonable costs, as reflected in gasoline taxes, tolls, other roadway fees, and other revenue sources devoted to maintaining performance.

Members of the bridge community such as designers, construction engineers, inspectors, maintenance engineers, and bridge management personnel who are responsible for maintaining performance must be able to properly and effectively evaluate bridge performance in more precise and targeted manners. They must break down bridge performance into very specific issues that can be adequately evaluated in terms of cause and effect. This will allow them to identify actions or programs that will ensure a high level of performance at a reasonable cost. This report describes a breakdown of bridge performance issues into four categories of performance: structural condition (for durability and serviceability), functionality (safety and traffic capacity), structural integrity (for safety and stability), and risk and costs to the user and to the agency. This breakdown helps isolate the most critical aspects of bridge performance and provides the basis for long-term research studies that will improve understanding of these issues.

Making a proper evaluation of bridge performance can be a formidable task given the many factors that can govern performance under different circumstances. The LTBP Program is being implemented to identify the most critical aspects of bridge performance and conduct studies that will result in the high-quality data necessary to better understand how multiple, variable factors govern those aspects of performance. This effort should ultimately lead to the ability to implement policies and actions for bridge programs that will improve performance and extend the life of bridges at minimum cost.

APPENDIX. DETAILS OF THE HIGHWAY BRIDGE INFRASTRUCTURE

Each bridge in the 2011 NBI is described by an extensive set of characteristics, parameters, and operating conditions, all of which have some impact on some aspect of the performance of the bridge. Table 9 provides a list of the most important of these items.

In addition, the age of the bridge is an important contributing factor to the current and future performance of the bridge. However, the simple age in years does not represent a precise measure of the impact of age on performance. The chronological age does not accurately reveal important knowledge about cumulative degradation of material properties, cumulative amount of damage from live loads, and past maintenance and repair history. The average age of all NHS bridges in the NBI is 36.3 years; the average age of all non-NHS bridges is 42.3 years; and the average age of all bridges is 41.0 years. Figure 1 in section 1 shows the diversity in age of bridges with a histogram of bridges still in service that were built within 5-year periods.

Other critical factors affecting bridge performance are the type, frequency, and effectiveness of preservation, maintenance, repair, and rehabilitation actions performed on bridges by the owner or entity charged with maintenance responsibility. Bridges on public highways are owned by a variety of agency types at different levels of government and by railroads, toll authorities, and other private entities. Table 2 in section 1 shows the number of different types of entities that have maintenance responsibilities for bridges on public highways.

Based on the 2011 NBI data, the bridge infrastructure in the United States can be further described by table 10, table 11, and figure 50.

Table 9. Diversity of Bridge Characteristics, Parameters, and Operating Conditions. [1]

Item	NBI Item Numbers
Kind of material, main span and/or approach span	43A, 43A
Structure type, main span and/or approach span	43B, 44B
Horizontal geometry and skew	47, 51, 52, 55A, 55B, 56
Vertical clearances over and under the bridge	53A, 54A, 54B
Design load	31
Bridge posting	70
Deck structure type	107
Wearing surface	108A
Membrane	108B
Protective system	108C
Type of joints and bearings	Data not available in NBI
Type of foundation	Data not available in NBI
Local environment	Data not available in NBI
Local climate patterns	Data not available in NBI
Maintenance, repair, and rehabilitation history	Data not available in NBI
Permit loads history	Data not available in NBI
Annual ADT and truck traffic	29, 109
Safety features	33, 36
Maintenance responsibility, owner	21, 22
Functional class of inventory route	26
Channel and channel protection	61
Critical feature inspection	92A, 92B, 92C
Scour critical	113

Table 10. Bridges by Functional Classification Weighted by Number, ADT, and Deck Area. [2]

Functional Class	Number	Percent	Deck Area (ft^2)	Percent
Rural, interstate	20,434	4.32	261,976,842	7.35
Rural, other arterial	51,304	10.84	540,109,441	15.16
Rural, collector	104,701	22.13	446,633,350	12.53
Rural, local	173,573	36.68	339,994,461	9.54
Subtotal, rural	350,012	73.97	1,588,714,083	44.58
Urban, interstate	26,774	5.66	697,385,694	19.57
Urban, other arterial	59,782	12.63	1,031,374,442	28.94
Urban, collector	14,812	3.13	117,799,536	3.31
Urban, local	21,785	4.60	128,570,034	3.61
Subtotal, urban	123,153	26.03	1,975,129,705	55.42
Total, rural and urban	473,165	100.00	3,563,843,788	100.00

Note: Table does not include culverts and tunnels.

Table 11. Numbers and Percentages of Bridges and Deck Area by Owner [2]

Bridge Owner	Number of Bridges	Percent of All Bridges	Deck Area (ft^2)	Percent of All Deck Area
State highway agency	214,058	45.23	2,589,978,541	72.66
State park, forest, or reservation agency	884	0.19	1,944,059	0.05
Other State agencies	1,080	0.23	10,281,397	0.29
State toll authority	6,861	1.45	127,080,287	3.57
Total, State Bridges	222,883	47.09	2,729,284,284	76.57
Other Federal agencies (not listed below)	52	0.01	1,308,741	0.04
Indian tribal government	1	0.00	312	0.00
Bureau of Indian Affairs	704	0.15	2,291,701	0.06
Bureau of Fish and Wildlife	278	0.06	371,064	0.01
U.S. Forest Service	3,579	0.76	4,647,426	0.13
National Park Service	1,150	0.24	6,227,557	0.17
Tennessee Valley Authority	35	0.01	1,072,268	0.03
Bureau of Land Management	1	0.00	1,518	0.00
Bureau of Reclamation	321	0.07	797,735	0.02
Corps of engineers (Civil)	441	0.09	5,438,014	0.15
Corps of engineers (Military)	17	0.00	659,440	0.02
Air Force	24	0.01	21,765	0.00
Navy/Marines	151	0.03	1,260,518	0.04
Army	556	0.12	2,149,574	0.06
National Aeronautics and Space Administration	0	0.00	0	0.00
Metropolitan Washington Airports Service	23	0.00	268,839	0.01
Total, Federal bridges	7,333	1.55	26,516,484	0.74
County highway agency	187,902	39.70	464,876,596	13.04
Town or township highway agency	23,563	4.98	41,438,795	1.16
City or municipal highway agency	27,998	5.92	235,181,884	6.60
Local park, forest, or reservation agency	64	0.01	139,285	0.00
Other local agencies	1,162	0.25	15,556,768	0.44
Local toll authority	582	0.12	37,121,035	1.04
Total, local bridges	241,271	50.98	794,314,352	22.28
Private (other than railroad)	433	0.09	6,549,226	0.18
Railroad	896	0.19	4,320,332	0.12
Total, private/railroad bridges	1,329	0.28	10,869,440	0.30
Unknown	363	0.08	1,901,929	0.05
Unclassified	89	0.02	1,597,300	0.04
Total, unknown/unclassified bridges	452	0.10	3,499,229	0.10

Note: Table does not include culverts and tunnels.

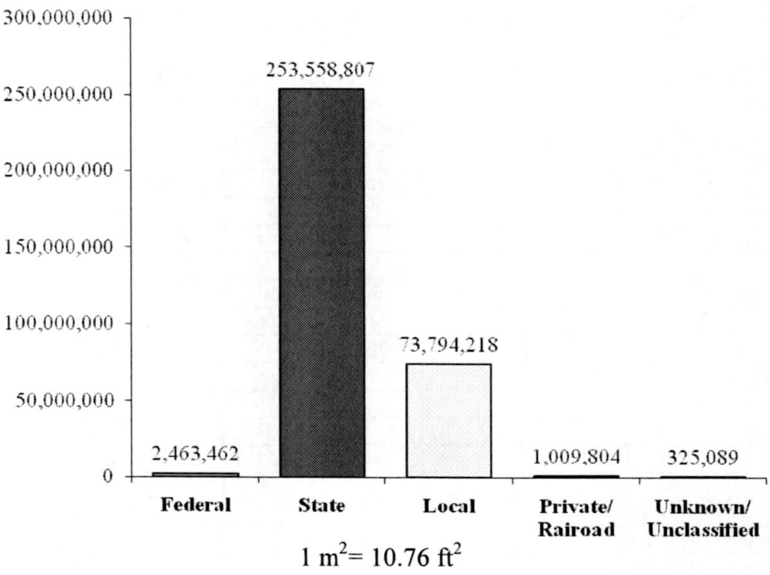

Figure 50. Graph. Deck Area (m2) of Bridges by Owners [2].

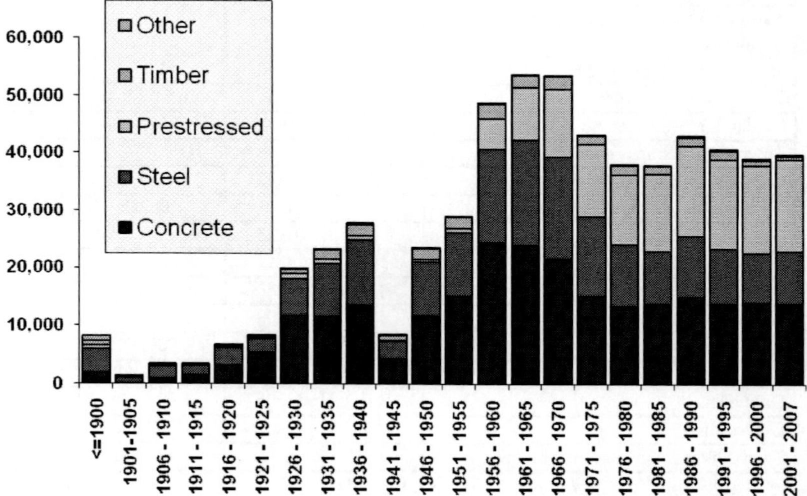

Figure 51. Graph. Numbers of Bridges Built in 5-Year Periods by Material [2].

Table 12. Design Type of Main Span (NBI Item Number 43B) [2]

Design Type	Number of Bridges	Design Type	Number of Bridges
Slab	80,239	Arch—thru	377
Stringer/multibeam girder	247,902	Suspension	97
Girder and floorbeam system	7,415	Stayed girder	46
Tee beam	35,749	Moveable lift	189
Box beam or girders—multiple	50,510	Moveable bascule	457
Box beam or girders—single or spread	9,201	Moveable swing	202
Frame (except frame culverts)	5,331	Tunnel	40
Orthotropic	472	Culverts	132,110
Truss—deck	465	Mixed types	22
Truss—thru	10,601	Segmental box girder	259
Arch—deck	6,794	Channel beam	14,442

Table 13. Kind of Material Main Span (NBI Item Number 43A) [2]

Design Type	Number of Bridges	Design Type	Number of Bridges
Concrete	176,800	Prestressed concrete continuous	22,733
Concrete continuous	74,796	Wood or timber	23,461
Steel	133,839	Masonry	1,706
Steel continuous	49,860	Aluminum, wrought or cast iron	1,470
Prestressed concrete	119,789	Moveable bascule	644

Note: Table does not include culverts and tunnels.

Table 14. Numbers of FO Bridges and Deck Area of FO Bridges by Owner [2]

Bridge Owner	Number of Bridges	Percent of All Bridges	Deck Area (ft^2)	Percent of All Deck Area
State highway agency	40,502	49.87	433,803,372	67.74
State park, forest, or reservation agency	190	0.23	471,169	0.07
Other State agencies	346	0.43	3,543,619	0.55
State toll authority	2,409	2.97	37,278,080	5.82
Total, State bridges	43,447	53.49	475,096,240	74.18
Total, Federal bridges	1,423	1.75	4,908,580	0.77
County highway agency	23,409	28.82	51,244,157	8.00
Town or township highway agency	2,424	2.98	3,551,628	0.55
City or municipal highway agency	8,160	10.05	70,987,020	11.08
Local park, forest, or reservation agency	22	0.03	51,914	0.01
Other local agencies	192	0.24	9,887,168	1.54
Local toll authority	190	0.23	15,963,805	2.49
Total, local bridges	35,820	44.10	156,594,273	24.45
Private (other than railroad)	110	0.14	1,920,572	0.30
Railroad	278	0.34	1,344,412	0.21
Total, private/railroad bridges	388	0.48	3,264,985	0.51
Unknown	141	0.17	559,433	0.09
Unclassified	0	0.00	0	0.00
Total, unknown/unclassified bridges	141	0.17	559,433	0.09

Note: Table does not include culverts and tunnels.

Table 15. Numbers of SD Bridges and Deck Area of SD Bridges by Owner [2]

Bridge Owner	Number of Bridges	Percent of All Bridges	Deck Area (ft^2)	Percent of All Deck Area
State highway agency	23,103	31.86	224,635,480	66.71
State park, forest, or reservation agency	161	0.22	256,558	0.08
Other State agencies	204	0.28	1,502,050	0.45
State toll authority	308	0.42	5,154,697	1.53
Total, State bridges	23,776	32.78	231,548,785	68.76
Total, Federal bridges	684	0.94	2,184,912	0.65
County highway agency	38,958	53.72	59,812,887	17.76
Town or township highway agency	3,972	5.48	4,603,498	1.37
City or municipal highway agency	4,391	6.05	31,643,905	9.40
Local park, forest, or reservation agency	12	0.02	24,273	0.01
Other local agencies	112	0.15	718,190	0.21
Local toll authority	38	0.05	3,610,226	1.07
Total, local bridges	47,483	65.48	100,412,979	29.82
Private (other than railroad)	61	0.08	479,543	0.14
Railroad	439	0.61	1,747,295	0.52
Total, private/railroad bridges	500	0.69	2,226,838	0.66
Unknown	78	0.11	382,011	0.11
Unclassified	0	0.00	0	0.00
Total, unknown/unclassified bridges	78	0.11	382,011	0.11

Note: Table does not include culverts and tunnels.

These tables and chart show that while ownership of the vast majority bridges by number is roughly split between State- and local-level agencies, the State-level agencies are responsible for almost 3.5 times as much deck area of bridges.

Table 12, table 13, and figure 51 show the variety of bridge types and materials that make up the Nation's bridge inventory.

Table 14 and table 15 show the number and deck area of bridges classified as FO and SD by ownership category.

REFERENCES

[1] Office of Engineering (1995). *Recording and Coding Guide for the Structure Inventory and Appraisal of the Nation's Bridges*, Report No. FHWA-PD-96-001, Federal Highway Administration, Washington, DC.

[2] Federal Highway Administration (2011). *National Bridge Inventory*, 2011 data, Washington, DC. Accessed online: August 7, 2012. (http://www.fhwa.dot.gov/bridge/nbi.htm)

[3] Slater, R.E. (1996). "The National Highway System: A Commitment to the Nation's Future." *Public Roads,* Vol. 59, No. 4, Federal Highway Administration, Washington, DC.

[4] Shepard, R.W. and Johnson, M.B. (1999). "California Bridge Health Index." Presented at the 8th International Bridge Management Conference, Transportation Research Board, Washington, DC.

[5] Office of International Programs (2005). *Bridge Preservation in Europe and South Africa*, Report No. FHWA-PL-05-002, Federal Highway Administration, Washington, DC.

[6] Frangopol, D.M., Lin, K-Y., and Estes, A.C. (1997). "Life-Cycle Cost Design of Deteriorating Structures." *Journal of Structural Engineering*, Vol. 123, No. 10, pp. 1390–1401, American Society of Civil Engineers, Reston, VA.

[7] Frangopol, D.M., Kong, J.S., and Gharaibeh, E.S. (2001). "Reliability-Based Life-Cycle Management of Highway Bridges." *Journal of Computing in Civil Engineering*, Vol. 15, No. 1, pp. 27–34, American Society of Civil Engineers, Reston, VA.

[8] Estes, A.C. and Frangopol, D.M. (2005). "Chapter 36: Life Cycle Evaluation and Condition Assessment of Structures." *Handbook of Structural Engineering*, Second Edition, W.F. Chen and E.M. Lui (eds.), CRC Press, Boca Raton, FL.

[9] Ang, A.H-S. and DeLeon, D. (2005). "Modeling and Analysis of

Uncertainties for Risk- Informed Decisions in Infrastructures Engineering." *Structure and Infrastructure Engineering*, Vol. 1, Issue 1, pp. 19–31, Taylor & Francis, Oxford, UK.

[10] Ellingwood, B.R. (2005). "Risk-Informed Condition Assessment of Civil Infrastructure: State of Practice and Research Issues." *Structure and Infrastructure Engineering,* Vol. 1, No. 1, pp. 7–18, Taylor & Francis, Oxford, UK.

[11] Frangopol, D.M. and Liu, M. (2007). "Maintenance and Management of Civil Infrastructure Based on Condition, Safety, Optimization, and Life-Cycle Cost." *Structure and Infrastructure Engineering*, Vol. 3, No. 1, pp. 29–41, Taylor & Francis, Oxford, UK.

[12] Frangopol, D.M. (2011). "Life-Cycle Performance, Management, and Optimization of Structural Systems under Uncertainty: Accomplishments and Challenges." *Structure and Infrastructure Engineering*, Vol. 7, No. 6, pp. 389–413, Taylor & Francis, Oxford, UK.

[13] American Association of State Highway and Transportation Officials (2011). *AASHTO Guide Manual for Bridge Element Inspection*, First Edition, Washington, DC.

[14] American Association of State Highway and Transportation Officials (2007). *AASHTO LRFD Bridge Design Specifications*, Fourth Edition, Washington, DC.

[15] Enright, M.P. and Frangopol, D.M. (1998). "Service-Life Prediction of Deteriorating Concrete Bridges." *Journal of Structural Engineering*, Vol. 124, No. 3, pp. 309–317, American Society of Civil Engineers, Reston, VA.

[16] Enright, M.P. and Frangopol, D.M. (1998). "Failure Time Prediction of Deteriorating Fail- Safe Structures." *Journal of Structural Engineering*, Vol. 124, No. 12, pp. 1448–1457, American Society of Civil Engineers, Reston, VA.

[17] Frangopol, D.M. and Messervey, T.B. (2009). "Life-Cycle Cost and Performance Prediction: Role of Structural Health Monitoring." *Frontier Technologies for Infrastructures Engineering*, S-S. Chen and A.H-S. Ang (eds.), CRC Press, Boca Raton, FL.

[18] Frangopol, D.M. and Messervey, T.B. (2009). "Maintenance Principles for Civil Structures." *Encyclopedia of Structural Health Monitoring*, Vol. 4, C. Boller, F-K. Chang, and Y. Fujino (eds.), John Wiley & Sons Ltd, Chicester, UK.

[19] American Association of State Highway and Transportation Officials (2002). *AASHTO Standard Specifications for Highway Bridges*, 17th Edition, Washington, DC.

[20] Cohn, M.Z. (1980). *Nonlinear Design of Concrete Structures: Problems and Prospects*, University of Waterloo Press, Ontario, Canada.

[21] Ang, A.H-S. and Tang, W.H. (2007). *Probability Concepts in Engineering. Emphasis on Applications to Civil and Environmental Engineering*, Second Edition, John Wiley & Sons, Hoboken, NJ.

[22] Mori, Y. and Ellingwood, B. (1993). "Reliability-Based Service-Life Assessment of Aging Concrete Structures." *Journal of Structural Engineering*, Vol. 119, No. 5, pp. 1600–1621, American Society of Civil Engineers, Reston, VA.

[23] Enright, M.P. and Frangopol, D.M. (1998). "Service-Life Prediction of Deteriorating Concrete Bridges." *Journal of Structural Engineering*, Vol. 124, No. 3, pp. 309–317, American Society of Civil Engineers, Reston, VA.

[24] Enright, M.P. and Frangopol, D.M. (1998). "Failure Time Prediction of Deteriorating Fail- Safe Structures." *Journal of Structural Engineering*, Vol. 124, No. 12, pp. 1448–1457, American Society of Civil Engineers, Reston, VA.

[25] Frangopol, D.M. and Curley, J.P. (1987). "Effects of Damage and Redundancy on Structural Reliability." *Journal of Structural Engineering*, Vol. 113, No. 7, pp. 1533–1549, American Society of Civil Engineers, Reston, VA.

[26] Frangopol, D.M. and Klisinski, M. (1989). "Material Behavior and Optimum Design of Structural Systems." *Journal of Structural Engineering*, Vol. 115, No. 5, pp. 1054–1075, American Society of Civil Engineers, Reston, VA.

[27] Frangopol, D.M. and Klisinski, M. (1989). "Weight-Strength-Redundancy Interaction in Optimum Design of Three-Dimensional Brittle-Ductile Trusses." *Computers and Structures*, Vol. 31, Issue 5, pp. 775–787, Pergamon Press, Oxford, UK.

[28] Frangopol, D.M. and Klisinski, M. (1992). "Design for Safety, Serviceability and Damage Tolerability." *Designing Concrete Structures for Serviceability and Safety*, Document No. SP133-12, American Concrete Institute, Farmington Hills, MI.

[29] Frangopol, D.M., Klisinski, M., and Lin K.Y. (1996). "Incorporating Damage Control in Structural Design." *Building an International Community of Structural Engineers*, Vol. 1, pp. 598–605, American Society of Civil Engineers, Reston, VA.

[30] Fu, G. (1987). *Modeling of Lifetime Structural System Reliability*, Report No. 87-9, Department of Civil Engineering, Case Western Reserve University, Cleveland, OH.

[31] Frangopol, D.M. and Nakib, R. (1991). "Redundancy in Highway Bridges." *Engineering Journal*, Vol. 28, No. 1, pp. 45–50, American Institute of Steel Construction, Chicago, IL.

[32] Frangopol, D.M. and Okasha, N.M. (2008). "Life-Cycle Performance and Redundancy of Structures." *Proceedings of the Sixth International Probabilistic Workshop*, pp. 1–14, Darmstadt, Germany.

[33] Okasha, N.M. and Frangopol, D.M. (2010). "Time-Variant Redundancy of Structural Systems." *Structure and Infrastructure Engineering*, Vol. 6, Nos. 1–2, pp. 279–301, Taylor & Francis, Oxford, UK.

[34] Lind, N.C. (1994). "A Measure of Vulnerability and Damage Tolerance." *Reliability Engineering and System Safety*, Vol. 43, No. 1, pp. 1–6, Elsevier, Amsterdam, Netherlands.

[35] Ghosn, M., Moses, F., and Frangopol, D.M. (2010). "Redundancy and Robustness of Highway Bridge Superstructures and Substructures." *Structure and Infrastructure Engineering*, Vol. 6, Nos. 1–2, pp. 257–278, Taylor & Francis, Oxford, UK.

[36] Tierney, K. and Bruneau, M. (2007). "Conceptualizing and Measuring Resilience." *TR News*, No. 250, pp. 14–17, Transportation Research Board, Washington, DC.

INDEX

D

E

F

N

O

P

Q

R

W

Y